R.A.S. No. 245.   Saturn, 1911, Nov. 19.   E. E. Barnard.

R.A.S. 291.   Eclipse, 1919 May 29.   Sobral, Brazil.   R.O.G. Expd.   Ap. 4 in., *f.l.* 228 in.   Exp. 28s.

[*Frontispiece.*

AN EASY OUTLINE OF ASTRONOMY

BY

M. DAVIDSON, D.Sc., F.R.A.S.

LONDON :

WATTS & CO.,

5 & 6 JOHNSON'S COURT, FLEET STREET, E.C.4

First published, 1943
Second Impression (revised), Dec., 1943
Third impression (slightly revised) 1946

THIS BOOK IS PRODUCED IN COMPLETE
CONFORMITY WITH THE
AUTHORISED ECONOMY STANDARDS

Printed and Published in Great Britain by C. A. Watts & Co. Limited,
5 & 6 Johnson's Court, Fleet Street, London, E.C.4

PREFACE

THIS work is intended for those who have no knowledge of astronomy, and has been designed to present the subject in the simplest language. For this reason technical terms and mathematical formulae have been avoided, as far as possible, and it is hoped that this simple outline will prove a stepping-stone to the more serious study of astronomy. The black-out has been responsible for arousing a considerable amount of interest in the heavenly bodies; and those who had given no thought to the wonders of the heavens in pre-war days are now finding how fascinating the subject can be.

The present work is confined to a description of the movements, dimensions, masses, and composition of the heavenly bodies, but in its limited scope it has been impossible to deal with the identification of the constellations and stars. Those who wish to pursue this side of the subject will find a short list of star charts at the end of the book, and any of these will enable them, with a little perseverance, to recognize the chief constellations and stars.

I wish to express my thanks to Sir Harold S. Jones, the Astronomer Royal, for permission to reproduce the photograph of the eclipse of the sun on May 29, 1919. I am also very grateful to Mr. F. J. Sellers, Editor of *The Journal of the British Astronomical Association*, for his assistance with the preparation of the diagrams. Finally, I wish to record my gratitude to the publishers, and especially to Mr. A. G. Whyte, for their interest in the preparation of the book and for numerous suggestions regarding its presentation.

Just before revising the text for the third impression my work *From Atoms to Stars* (Hutchinson's Scientific and Technical Publications) appeared. This deals much more fully with many of the points merely referred to in the present volume.

M. DAVIDSON.

August, 1943.

CONTENTS

INTRODUCTION

ASTRONOMY is the oldest of all sciences, and has rightly been called the Queen of the Sciences. It is a subject that has attracted the attention of men from the earliest times, though there are no records which tell us of the dawn of astronomy. We know for certain that the early inhabitants of the earth not only regarded the heavenly bodies with wonder and awe; they even built temples in their honour and worshipped a number of them as deities.

Those who make no claim to be astronomers are aware of the apparent daily motion of the sun across the heavens from east to west, and it is only necessary to observe the stars for about an hour any night to see that they also share in this apparent motion. Like the sun, stars are seen rising in the east and slowly mounting higher above the horizon, while others near the west are observed to move nearer and nearer to the horizon and finally to disappear. It is not surprising that people once believed the earth was the centre of the universe and that all the heavenly bodies moved round it, completing their course in 24 hours. We know now that the apparent motion of the stars is due to the spinning motion of the earth, which is explained more fully in the first chapter. It is remarkable that there was a time when people who dared to suggest that the earth moved might lose their lives. It is just 310 years ago since Galileo was condemned for maintaining that the earth moved, and he saved his life only by renouncing his " errors and heresies." Fortunately we now live in more tolerant days when there is nothing to prevent men of science from advocating their views even if they are erroneous. In spite of the enormous progress in astronomy, some people still believe that the earth is flat and that the heavenly bodies are all moving round it. Such people can become very tiresome if one tries to convince them of their errors. Some time ago one of them wrote to me and advanced a number of " arguments " disproving the motion of the earth. I tried to deal with his difficulties, but without any success, and finally he wrote a very rude letter, in which he accused me of being an Atheist. As evidence of this he quoted Psalm civ, 5: " Who laid the foundations of the earth, that it should not be moved for ever." The Royal Observatory, Greenwich, still receives communications from flat-earthers who supply " proofs " that the earth is fixed in space and that all the heavenly bodies are moving round it as centre.

You may have noticed at a railway station that when your train is moving out of the station, another train near by seems to be moving in the opposite direction. In fact, you may be uncertain for a time whether your train is moving or whether it is the other train that is in motion. It is the same with the earth and the stars. We think that the stars are moving from east to west, but things would appear the same if the stars were standing still and the earth was turning from west to east. There are certain experiments which prove conclusively that the earth is turning in this way, but you must wait until you have read more advanced text-books on astronomy before you can understand these experiments.

The following argument will prove convincing in the meantime. If the heavenly bodies are all moving round the earth as a centre, then those which are far away must have enormous speeds compared with those which are nearer. Now we know that the stars are situated at different distances from the earth (this point will be dealt with later), and hence we must conclude that their speeds are so adjusted that those which are far off must move millions of times faster than those which are nearest to us, in order that they may all complete their journeys in the same time. How much simpler it is to accept the fact that the earth is spinning round once in a day and that the motions of the heavenly bodies are only apparent, and are caused by the motion of the earth in the opposite direction.

When you look at the sky on a clear night you may see the moon and perhaps two or three planets, but by far the most numerous of the heavenly bodies which are visible are the stars. It is important that you should know the distinction between a star and a planet, not only by their appearances, but also by the great difference in their condition.   The best way to do so is to remember that the sun is a typical star and the earth a typical planet. Astronomers can understand very much better what they observe on the stars if they first study the sun.   They can

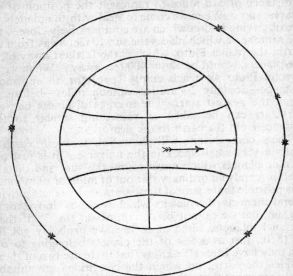

FIG. 1.—As the earth rotates in the direction shown by the arrow the heavenly bodies appear to move in the opposite direction.

also understand the planets better when they have made a careful study of the earth—not only of its ordinary features, such as mountains, oceans, seas, etc., but also of its atmosphere, and even of the conditions far down in its interior.   When you look at a star, remember that you are looking at a body which is similar to the sun. It is intensely hot and bright, and because of its brightness we can see it.   Although many of the stars are very much hotter than the sun, we derive no advantage from their heat, because of their enormous distances in comparison with the distance of the sun. The nearest star is about 270 thousand times as far away from the earth as the sun is.

A useful comparison can be obtained as follows : Imagine that the sun is situated at a distance of 10 feet from the earth; a model on the same scale would require a distance of 510 miles to represent the position of the nearest star. When we come to deal with the planets we are observing bodies which are comparatively close. On the same scale which places the sun 10 feet away from the earth, the planet Pluto, which is the farthest away of all the planets, would be about 400 feet away. When we go beyond Pluto we reach empty space for an enormous distance—actually 26 million million miles—before we reach the nearest star. The average distances between the stars can be taken as something similar to this distance from the earth to the nearest star.

These comparisons will show you that our earth is really a very small speck in the universe. It is just one of nine planets which move round the sun, and the sun itself is merely an ordinary star out of millions of millions. It is possible that many of these stars have planets moving round them as the planets which we know move round the sun, but we cannot be certain about this.* If there are such planets, some of them have probably got life on them, just as a few of the planets belonging to our sun may have life. We know that there are two of them, besides the earth, on which the conditions are suitable for life in some form, though of course it is most unlikely that the highest form of life on them resembles human beings. Something will be said in a later chapter on this point, and we can leave it for the present.

Before giving a more detailed description of the stars and each planet, it is advisable that the reader should

---

* While reading the proofs it was announced that a planetary companion to the double star 61 Cygni had been discovered. The body is estimated to have a mass sixteen times that of Jupiter, and hence is large for a planet. It is the first discovery of a planetary body outside the solar system. This new planet has not been seen, as it is too small and faint to be observed even in the most powerful telescope. Its presence is known from its disturbance to the motion of the components of the double star 61 Cygni.

obtain a general view of the system of stars to which we belong. At any place it is possible to see about 3,000 stars with the naked eye, and as we can see only half the total number of stars at any place, the number which is within the range of the naked eye is about 6,000. Of course this is only a very minute fraction of the number of stars which can be observed with a moderate-sized telescope or which is within the range of the astronomer's photographic equipment. All the stars which you see belong to what is known as the *Galaxy*, which contains anything from 30 thousand million stars to twice that number (up to the present the number has not been determined with great accuracy). You will see that 6,000 stars out of this enormous number is very small; nevertheless these 6,000 can be taken as typical of those in the whole galaxy. The Milky Way is another name for the galaxy,* and you have probably seen it on a clear night as a luminous girdle around the heavens. When you remember the number of stars in it, it is not surprising that it appears luminous. Of course they are too far off to be separately visible to the naked eye, but the combined light of these millions of stars produces the faint glow.

In some places you may have noticed rifts in the girdle. These are not due to the absence of stars, but are caused by dust-clouds which cut off the light from us, just as a sandstorm cuts off the light of the sun, or a cloud makes the sun look darker. The galaxy is only one of hundreds of thousands of systems of stars which are separated from one another by vast distances, but at present we shall confine our attention to our own galaxy. A model of this is shown in Fig. 11, Chapter VIII. It is seen from this that the galaxy is bun-shaped, its thickness at the centre being about one-sixth of its diameter *AB*. The thickness diminishes rapidly from the centre to the outside of the system, so that, taken on the whole, the

---

* Galaxy and Milky Way are often used to mean our entire system of stars, but strictly speaking " Milky Way " should be restricted to the portion of the galaxy which shows as a white girdle across the heavens. The stars on either side of this girdle belong to our galaxy, but are not considered to form part of the Milky Way.

galaxy has a very flattened appearance. Among the
30 thousand million stars in the galaxy there is one
which is important for our purpose, because from it we
derive our heat and light, and without it all life on earth
would perish. This star is called the sun. It is not
situated at the centre of the galaxy, as astronomers once
believed, but is very much to one side, and is marked X
in the diagram.

If there are astronomers on planets belonging to any
of the other stars, they would be able to see the sun with
or without their telescopes, and they would probably
find out that it was not a very conspicuous star, being
very much fainter and smaller than some, though
brighter and larger than many others. None of the
planets belonging to the sun would be visible even with
the most powerful telescopes, and the astronomers might
speculate on the possibility of planets existing in association with the sun, just as astronomers on the earth speculate on the possibility of planets being associated with
the various stars. Our most powerful telescopes fail to
detect such planets, and probably will always fail to
detect them, even if they are much larger than any of the
sun's planets. Not only does the small size of a planet
prevent us from seeing it at such vast distances; more
important is the fact that a planet has no light of its own,
and so does not shine like a star. When we see the
planets belonging to the sun, we do so by the reflected
light of the sun. This is very faint in comparison with the
light which is directly sent out by a shining body like a
star, but as the planets are comparatively near us we are
able to see some of them by aid of this reflected light,
even with the naked eye.

From this brief survey of the heavenly bodies you will
be able to follow the descriptions of a number of separate
planets and stars in the following chapters.

Try to keep in mind the important fact that the earth
is not the centre of the universe, but is just an ordinary
planet which is spinning round once every day and
which is also moving round the sun in a curve that is
nearly but not quite a circle, completing one revolution
in a year.

CHAPTER I

THE EARTH

Dimensions and Shape of the Earth. From the point of view of human beings the earth is the most important of all the planets. Probably if intelligent beings exist on other planets they would not be prepared to share this view, as their own abode would naturally be the most important for them. It is fitting that we should start with a description of the scene of our own life and activities, and a knowledge of the conditions of the earth will assist us in understanding those which exist on other planets.

The earth is not a perfect sphere, though for many practical purposes it can be taken as a sphere. It is slightly flattened at the poles, like an orange, and it bulges at the equator. This peculiar shape is due to its rotation on its axis, about which more will be said later. A line joining the north and south poles, known as the polar axis or polar diameter, is 7,900 miles in length, and a line at the equator drawn perpendicular to the polar axis, known as the equatorial diameter, is 7,927 miles in length.

It is difficult for us to observe the curvature of the earth, because this is so very slight within our usual limited range of vision. One of the simplest proofs of the rotundity of the earth is the gradual disappearance of a ship as it recedes from land, the masts and funnel remaining visible after the hull has vanished. If the sea is perfectly still and the eye is placed close to the surface of the water, it is possible to detect the curvature of the earth over a fairly short distance. The drop below the line of sight is 8 inches for the first mile, 32 inches for 2 miles, 72 inches for 3 miles, and so on. If the drop for 8 miles is required, it is only necessary to multiply 64, which is the square of 8, by 8 inches, and the result, 512 inches, is the required drop.

Another proof that the earth is round is obtained from

eclipses of the moon. An eclipse of the moon is caused by the earth casting its shadow on the moon; the sun, earth and moon being in a line. As the shadow creeps across the face of the moon it is seen to be curved, just as we should expect from a round body like the earth. I have not yet discovered how the flat-earthers explain eclipses of either the sun or of the moon, but as they are highly ingenious in their methods of explanation, I have no doubt that they are able to account for these phenomena, at least to their own satisfaction.

The Earth's Interior and Atmosphere. We have very little direct knowledge about the interior of the earth beyond a few miles in depth, but there is a certain amount of indirect evidence derived especially from a study of earthquakes. This subject is too difficult to explain fully in an elementary treatise, and research on earthquakes, their causes, the speed with which the shocks travel through the earth, etc., is the work of specialists. One thing may be mentioned in connection with the conditions far down in the earth's interior. Earthquake shocks differ according to the nature of the material through which tremors pass. Thus, there is a great difference, which the experts can detect immediately, between tremors sent through a solid mass like a great collection of rocks and those sent through something in a liquid state. By studying these tremors we gather that if we could go down about 2,000 miles into the earth's interior we should come to iron in a molten state, but heavier than the iron which we use every day. From the centre of the earth, therefore, for 2,000 miles all round, there is this dense "central core," as it is called. Above the central core there is the crust which comes to within 40 or 50 miles from the surface, and this consists of rocks which are about four times as heavy as water, for the same bulk. On the top of this crust we have other rocks not quite as heavy as those underneath. (The average density * of the earth is about $5\frac{1}{2}$.) Above this last crust we come to the land, and oceans, and above these the atmosphere.

* By the density of any substance we mean its weight, if the weight of the same bulk of water is taken as the unit.

Like most of the planets, the earth has an atmosphere, though this differs very much from those which exist on other planets. It is scarcely necessary to say that life, as we know it, would be impossible without an atmosphere. Even if life could exist without an atmosphere it could last only a very short time, because the bombardment of the earth's surface by meteors would soon destroy it. These meteors, about which something will be said in Chapter V, are usually burnt up, owing to the heat produced by their friction with the atmosphere, at heights of 30 to 40 miles. They are moving with high speeds—varying from 10 to 45 miles a second—and become intensely hot by their collision with the molecules of the atmosphere. Because they become white-hot for a few seconds as they rush along, their sudden bright appearance has been responsible for the usual title, "shooting stars." Sometimes the larger meteors are not completely burnt up in the atmosphere and hence reach the earth; they are then known as "meteorites," and the study of these bodies has become a very important branch of science in recent years. As meteors are sometimes visible at heights of over 100 miles, we know that the atmosphere must extend to this distance, though it is very thin at such heights. Aurorae, known in the northern hemisphere as the "Northern Lights," have been observed at much greater heights still, from which it may be presumed that even there an atmosphere of some kind exists.

The atmosphere is important from more points of view than assisting with sustaining life. In comparatively recent times we have discovered that certain parts of it are responsible for reflecting radio waves. The "Kennelly-Heaviside layer," discovered in 1902 by Kennelly in the United States and Heaviside in England, is generally found at a height of about 65 miles, but its height may vary by as much as 20 miles on either side of this. This layer reflects radio waves, just as an ordinary mirror reflects light, and turns them back to the earth. Higher still is another layer, named the Appleton layer after its discoverer. Its height may be as much as 250 miles or as low as 90 miles, and it too reflects radio waves back to earth. There is a lower layer which is only about 30 miles above the surface of the earth, and this serves the useful purpose of reflecting the long waves back to earth, but it is not very effective in reflecting the short waves. The other layers referred to are able to do this much better.

Revolution and Rotation of the Earth. The earth revolves round the sun, completing its journey in a year and moving with a speed of about $18\frac{1}{2}$ miles a second. The path which it traces in its yearly motion is not a circle but an ellipse, and the same kind of curve is pursued by all the planets and comets in their revolution round the sun. There was a time when people thought that circular motion was " perfect," and this view was responsible for delaying important discoveries in connection with the movements of the planets. The Greek astronomers tried to represent the movements of the heavenly bodies by circles, and in the Middle Ages astronomers of note, like Copernicus and Tycho, retained circular motion in their explanations. Kepler tried to represent the motions of the planets on the same pattern, but finally gave up the attempt, and to his great delight discovered that the simplest of all oval curves— the ellipse—satisfied the observations of the positions of the planets. He announced his law about motion in an ellipse in 1609, and his book *The Harmony of the World*, which contains this and his other laws, was published nine years later.

An ellipse is easily traced on a sheet of paper with the aid of two pins and a piece of thread in the form of a loop. Insert the pins in the paper; place the thread over them and pull it tight with a pencil point, which can now be moved along the paper, tracing out an ellipse. When the pins are moved farther apart it will be noticed that the ellipse becomes flatter. If the pins are brought closer together the ellipse becomes more like a circle, and when the two pins merge into one on the same spot, the curve becomes a circle. The point marked by each pin is called the focus of the ellipse. A very elongated ellipse is shown in Fig. 9, Chapter V.

The sun is in one of the foci of the ellipse which the

earth—in common with all the other planets—describes in its orbital motion round the sun. For this reason no planet keeps a constant distance from the sun in its orbit, the name given to the course of a planet or any other body round the sun. No two planetary orbits are the same in shape, though all are ellipses. Some are more oval than others, and hence show a greater difference between the nearest and the farthest distances from the sun than when the shape of the ellipse is more like a circle. The small scale on which the diagrams are drawn shows the planetary orbits as circles.

In addition to the earth's revolution round the sun there is also its daily rotation, which gives us the succession of day and night. The difference between *revolution* and *rotation* should be carefully noticed, because astronomers use the words to denote two distinct movements. *Revolution* is motion around an outside centre, as when the earth or any other body moves round the sun, which is then the outside centre. *Rotation* is motion around an axis within the body itself. A good illustration can be arranged by means of an orange through the flattened ends of which a needle is inserted. If the orange is carried round a central body, say a lamp on a table, this movement is similar to the earth's orbital revolution round the sun. If it is spun like a top by turning the needle, this movement is similar to the earth's axial rotation. Of course there is no real axis through the earth corresponding to the needle in the orange; the earth's axis is the *imaginary* line which is drawn from pole to pole. As explained in the Introduction, the axial rotation of the earth from west to east is responsible for the apparent movement of the heavenly bodies in the opposite direction.

A very important effect of the axial rotation is that it keeps the earth's axis pointing in the same direction—almost exactly, in fact, towards the pole star, which is such a useful guide to sailors and explorers. If the earth did not spin like a top, its axis would not tend to maintain this steady position. In the well-known toy called a gyroscope—a special kind of top—we see the same forces at work. When we hold a spinning gyroscope in our hand we find it difficult to change the direction of the axle; the gyroscope seems to behave like a live thing, stubbornly resisting every effort to tilt the axle. At the same time we can easily carry the gyroscope round the room, so long as we do not try to make the axle point in a new direction. The reasons for this curious behaviour of a spinning mass cannot be given without a great deal of complicated mathematics, but any one can convince himself of the facts by playing with the toy. If the earth, which rotates once in 24 hours, does not seem to be spinning fast enough to behave like a gyroscope, we must remember that for its size it is really rotating very rapidly. A man standing at the equator is carried round at a speed of 1,000 miles an hour.

It should be added that the earth's axis does not keep *absolutely* in the same direction all the time. Like a top near the end of its spin, it " wobbles," and the axis has also a revolution of its own, but as the latter is slow and the former is very slight, we can leave them out of account when explaining how the behaviour of the earth accounts for the four seasons.

Explanation of the Seasons. It has already been explained that the earth revolves round the sun in an orbit which is an ellipse. If we imagine a line drawn from the centre of the sun to the centre of the earth, as the earth moves round the sun this line would sweep out a flat surface which astronomers call " the plane of the ecliptic." We have not yet said anything about the position of the earth's axis with reference to this plane. If the axis had been perpendicular to the plane of the ecliptic there would have been no seasons, as any particular place on the surface of the earth would get the same amount of heat and light each day all the year, except, of course, for cloud interference. There would be no change from long cold winter nights to long warm summer days. The earth's axis, however, is tilted at an angle of about $66\frac{1}{2}°$ to the plane of the ecliptic. Fig. 2 shows how this tilt gives us the seasons; the broken line represents the orbit of the earth. Readers will find the explanation easier to follow if they use a ball about 6 inches or more in diameter on which two marks are placed which correspond to P and Q, the north and south poles respectively.

Starting with the earth in the position 1, let the surface of the paper represent the plane of the ecliptic in which the sun S is situated. In position 1 let the axis PQ be inclined to the plane of the paper at an angle of $66\frac{1}{2}°$. It will be found, by using the model, that there will be hundreds of positions for the axis with this inclination. Thus P may be nearer to S than Q is, or Q may be

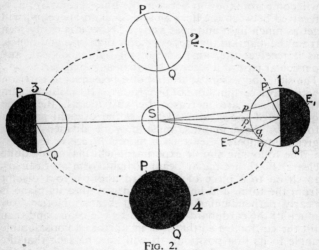

FIG. 2.

nearer to S than P is, and there are numerous intermediate positions, although the inclination of PQ to the plane of the paper remains at $66\frac{1}{2}°$. If, however, it is decided that P shall be as near to S as is possible, and therefore Q as far away from S as is possible, *there is only one position in which the axis PQ can be placed*; in this position the north pole, P, is tilted towards the sun, the south pole, Q, being tilted away from the sun, as far as possible in each case: 1 is the position which corresponds to midsummer on the northern hemisphere and to mid-winter on the southern hemisphere.

B

To show why a place in the northern hemisphere receives more heat at this time than a place at the same latitude in the southern hemisphere, draw a line from S to the centre of the earth, cutting the surface of the earth at p_1. EE_1 represents the equator and q is a point on the surface of the earth as far south of E as p_1 is north of E. From S draw a line Sp making a small angle—say 5°—with Sp_1, and another line Sq_1 making the same angle with Sq. It is clear that the same quantity of heat and light will come from the sun between the lines Sp and Sp_1 as is received between the lines Sq and Sq_1, and hence pp_1 will get as much light and heat as qq_1. Now it is easily seen from the diagram that qq_1 is greater than pp_1, and hence, as it receives the same amount of light and heat, any particular part of it receives less than an equal part in pp_1. The situation is similar to that of a beleaguered garrison with only a certain store of provisions. If the numbers in the garrison are increased by any means, the same quantity of food, rationed to last the original garrison for, say, three months, must now be divided among greater numbers; hence each man will receive less.

The following simple experiment will make this quite clear: Flash a circular beam of light from an ordinary torch on to a piece of cardboard about a foot distant from the torch and held in such a way that its plane is nearly perpendicular to the light beam. Notice how much of the cardboard is covered by the beam, and then tilt the cardboard so that it is inclined at a considerable angle to its first position. It will now be seen that the light from the torch covers a much larger surface of the cardboard than it did in the first position, so that any definite area, say a square inch of the cardboard, will receive less light than it did in the first instance. The same remark also applies to the heat given out by the torch, but as this is very small it would be impossible to measure it by ordinary methods. The principle is the same, however, for both heat and light, and it will now be clear why position 1 in the diagram corresponds to midsummer in the northern hemisphere and to midwinter in the southern hemisphere.

We follow the same principle in many familiar actions.

When we are reading we hold the page so that the light falls full upon it, and not at a low slant, as we find that we get more illumination in that position. When we are airing a garment in front of a fire we place the article so that the heat-rays fall vertically upon it.

It has already been pointed out that the axis PQ maintains a constant direction in space as the earth moves round the sun. In position 3, therefore, the axis points in the same direction as it did when the earth was at 1. This can be illustrated on the model by making PQ point to a top corner of the room as the model is carried round a central body which represents the sun. The circle in which it is moved should not be large, as otherwise pointing PQ to the corner of the room will not necessarily ensure that PQ remains nearly parallel to itself; in other words, pointing nearly in the same direction. In position 3 we have the exact opposite to that which we found at 1. The south pole, Q, is now tilted towards the sun, and the light and heat will fall more directly on the southern than they do on the northern hemisphere; Z corresponds to midsummer on the southern and to midwinter on the northern hemisphere. In the positions 2 and 4 the axis is not pointed either towards or away from the sun, and places at the same latitude in each hemisphere will then receive the rays from the sun at exactly the same angle. These positions correspond to the autumn and spring equinoxes respectively, when day and night are equal in length.

There is another factor which assists in causing the seasons. When the sun's rays fall obliquely on either hemisphere during its winter, they must pass through a greater thickness of air than they do when they strike the earth directly. The air, even when it is free from cloud, absorbs a certain amount of the sun's heat, and hence has some cooling effect of its own, quite apart from the lessened heat due to the rays falling obliquely. While this effect is not very large, it plays its part in producing the seasons.

Strange as it may seem at first sight, the distance of the earth from the sun has very little to do with the difference between summer and winter. In fact, in the northern hemisphere the earth is about three million miles *nearer* to the sun at midwinter than it is at midsummer. In the southern hemisphere, on the other hand, it is about three million miles nearer to the sun at midsummer than it is at midwinter. One might think that the summer at a certain place on the southern hemisphere should be hotter than a place in the same latitude on the northern hemisphere, but there are many other factors, such as height above sea-level, distribution of land and water, direction of prevailing winds, etc., which are also important in determining the temperature of a locality.

It will be seen from Fig. 2 that in position 1 the sun does not set in regions near *P*, the north pole, or rise in places near the south pole. The axial rotation on *PQ* does not take any locality between *P* and the portion shaded dark away from the light of the sun. Between 1 and 2 the continuously illuminated patch will slowly decrease, and at 2—the autumn equinox—the continuously illuminated part will not extend south of *P*. Between 2 and 3 *P* is tilted away from the sun and the axial rotation will not take *P* into the sunshine. At 3 there is perpetual darkness over a large portion of the regions surrounding *P*. When we speak of perpetual darkness we are leaving out of account the effects of twilight, which shortens the polar night very considerably.

Questions

1. If anyone told you that the earth is not round, what proofs would you give to show that it is round?

2. Give a brief account of what is believed to be the condition in the interior of the earth.

3. Describe the difference between the two motions, revolution and rotation. Which of them causes the succession of day and night?

4. How would you propose to arrange the inclination of the axis of the earth, or any planet, with regard to the plane of the ecliptic, to produce the following effects : (*a*) no changes in the seasons ; (*b*) the greatest possible changes in the seasons?

5. How much nearer are we to the sun at midwinter in the northern hemisphere than at midsummer? Is our winter much milder on this account than the winter of the southern hemisphere?

6. Explain why there are 6 months of darkness and 6 months of day at each pole in turn. Why do places at some distance from each pole get less than 6 months' continuous darkness and light?

7. Show from Fig. 2 how far north you would go to see the midnight sun at midsummer.

CHAPTER II

THE MOON

The Motion of the Moon. The moon is our nearest neighbour and, next to the sun, is the most useful for our needs. It is true that we do not depend on the moon for our very existence, as we do on the sun, but if the moon disappeared from the heavens we should miss her very much, for two reasons. First of all we should miss the moonlight, which, though not indispensable, yet renders great assistance to the denizens of this planet, especially in the winter months. Then we should miss her influence on the seas, because she is more important than the sun in producing the tides.

The moon depends on the earth for her motion in the heavens. We have only to notice how she moves farther east among the stars from the first appearance of the crescent until full moon to infer that she is revolving around the earth just as the earth is revolving around the sun. Perhaps this statement should be modified slightly to prevent ´misunderstandings. When one heavenly body revolves around another we are accustomed to think of the larger body remaining at rest. Actually both bodies revolve around their common centre of gravity. This principle can be illustrated very simply by attaching two weights to the end of a rigid rod, finding the position of the rod where it balances on a pivot (known as the centre of gravity of the system), and turning the rod on its pivot. This gives a rough picture of the revolution of the heavenly bodies, and if one body is much heavier than the other, the centre of gravity may be inside the heavier of the two. This happens in the case of the sun and the earth. The sun weighs more than 300,000 times as much as the earth, so that, in spite of the enormous distance between them, the centre of gravity of the sun–earth system is only about 300 miles from the centre of the sun—a very small distance for a body the diameter of which is 864,000 miles.

The moon's weight is less than one-eightieth of the earth's weight, and as her distance is about 240,000 miles, the centre of gravity of the earth–moon system, known as the "barycentre," is 3,000 miles from the centre of the earth. This result is obtained by dividing 240,000 by 80. As the earth's radius is 4,000 miles, the centre of gravity of the earth–moon system is about 1,000 miles below the earth's surface. It is interesting to know, therefore, that our planet, in addition to its revolution round the sun, is also revolving round a point 1,000 miles below its surface. The moon revolves round the earth, or rather round the barycentre, once in 27 days 7 hours, and hence the earth's centre also revolves round this barycentre in the same period. As the moon always turns the same face towards the earth, we never see her other side, but it probably resembles the face that we see.*

Phases of the Moon. The revolution of the moon round the earth is responsible for the "phases" through which she passes in a lunar month and with which everyone is familiar. The light which the moon sends to us is not her own, because, unlike the sun and stars, she has no light or heat. When we see her we do so because she reflects the light which the sun sends to her. If there were inhabitants on the moon they would see the earth in just the same way as we see our satellite—by the light of the sun reflected by the earth. Fig. 3 shows the reasons for the moon's phases, but readers are strongly advised to make a small model which will tell them more about the lunar phases than many pages of explanation. In the diagram E is the earth, S the sun, and M the moon in various positions in her orbital motion around the earth. In position 1 the sun, moon, and earth are nearly in a straight line, and the portion of the moon turned towards the earth is in the dark, so that she is invisible. It has already been pointed out that the light which we derive from the moon is merely reflected sunlight; at 1 this sunlight is reflected away from the

---

* This is not quite correct, as the moon has a slight libration—a rocking from side to side—in consequence of which we see 59 per cent of her whole surface.

earth. In position 3 the hemisphere of the moon which is illuminated by the sun is turned towards the earth, and the moon looks like a circular disc. In this case we speak of the moon being full. In intermediate positions, say at 2 or 4, the dark side is not turned towards us completely nor is the illuminated side fully turned towards us. From the diagram we see that half the illuminated hemisphere is visible to us; in this position we speak of the " half moon." Between 1 and 2 less than half the

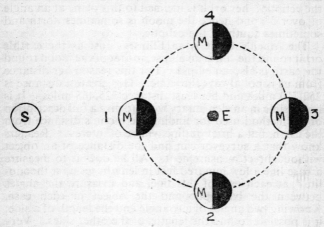

FIG. 3.

hemisphere will be visible, and very close to 1 a thin crescent will be seen which we describe as the " new moon." In positions after full moon the phases will repeat themselves, but the portions which we see as three-quarters moon, half moon, etc., will now appear in the reverse order, and the crescent part will be on a different edge of the moon.

The model can consist of two spheres—two billiard or tennis balls—to illustrate the phases, and the light can be supplied by an electric torch which represents the sun. The plane of the torch and the ball which is supposed to be the earth should not pass through the other ball which represents the moon. In fact, if the torch and the earth ball are on a table, the other ball should be a little above the level of the table. On moving the moon ball around the other one it will be seen how the illumination of a hemisphere appears to observers on the earth, according to the relative positions of sun, earth, and moon. The reason for placing the moon ball above the table, and not in the plane of the earth and sun, is that the moon does not revolve round the earth in the plane of the ecliptic; her orbit is inclined to this plane at an angle of over 5°, and hence the moon is sometimes north and sometimes south of the ecliptic.

The Moon's Orbit is an Ellipse. Just as the earth's orbit round the sun is an ellipse, so that of the moon round the earth is also an ellipse. For this reason her distance from us is not always the same. Her greatest distance is 252,710 miles and her least distance 225,463 miles. It is impossible in an elementary work to give a full description of the method used for finding the moon's distance from the earth, but a brief outline will be of interest. Readers know that a surveyor can find the distance of an object without direct measurement. All he does is to measure a base line a few hundred feet in length, set up a theodolite * at each end of this line, and measure the angles between the base line and the object in each case. Knowing two angles of a triangle and the length of a side, it is possible to find the lengths of the other sides. Very great accuracy can be obtained by this method; in fact it was used in preparing our Ordnance Maps.

The same principle is applied in measuring the distance of the moon from the earth. In this case, however, a base line of a few hundred feet would be of little use, as the moon is so far away. Astronomers used one a few thousand miles long connecting two observatories, one in each hemisphere, thus securing the longest base line which could be conveniently used. By measuring the distance of the moon at different times they found that her orbit is not circular, but an ellipse.

Size of the Moon. Knowing the moon's distance, it

* A theodolite is an instrument for measuring horizontal and vertical angles.

is easy to find her diameter. It is only necessary to measure the angle which her diameter " subtends " at the earth—that is, to find the angle between a line from the eye to the extreme left-hand edge of the full moon and a line from the eye to the extreme right-hand edge. A simple calculation then gives the moon's diameter, which is 2,160 miles. The angle which the moon subtends at the earth varies a little, owing to the moon's different distances from the earth.

There is a certain ambiguity in describing the distance of the moon from the earth. It does not mean the distance from the surface of the moon to the place where the observer on the earth is standing, but the distance between the centres of the earth and the moon.

Force of Gravity on the Moon's Surface. On the earth we know that a body falls 16 feet in one second, 64 feet in two seconds, and so on, but on the moon a body would fall only $2\frac{1}{2}$ feet in a second, 10 feet in two seconds, and so on. If you could throw a stone 60 feet high on the earth you could throw it over 360 feet high on the moon. Similarly, a man on the moon would have little difficulty in clearing a wall 25 feet high, or even higher. The reason is that the moon's attraction for bodies is only about one-sixth that of the earth's attraction, and this is due to the moon's weight being so much less than that of the earth. There are many small bodies like the minor planets, about which we shall say something later, the attractions of which are so small that if a man jumped upwards from one of them he would never return. The body would lose control of him, and he would just go out into space, perhaps to be captured in the course of myriads of years by another larger body.

The Moon Has no Atmosphere. The moon to-day has no atmosphere. If she ever possessed an atmosphere it has disappeared, and this is not surprising when the small force of gravity on the surface of the moon is considered. The molecules of our atmosphere are in constant motion, and if any of them in the upper regions of the atmosphere, where they would not be so likely to collide with other molecules, had ever a speed of 7 miles a second or more—which is the *velocity of escape* for the earth—they would move off into space, the earth losing control of them. If a big gun could project from the earth a shell with a velocity of 7 miles a second, the shell would not return. On the moon the corresponding figure is about 1½ miles a second, and if she ever had an atmosphere it is obvious that it would be lost much more easily than is the case with the earth.

Although Prof. W. H. Pickering thought he had detected signs of vegetation on certain parts of the moon, which would imply the presence of an atmosphere, English astronomers were unwilling to accept his results, and it is now generally agreed that the moon is utterly devoid of an atmosphere. Observations support this view. When the moon comes between us and a star, the star is hidden from us for a time and is said to be occulted. In every case of occultation by the moon the star vanishes suddenly and completely. If the moon had an atmosphere, some of the star's light, even when the star had just gone behind the moon, would be scattered by the atmosphere, and for this reason the star would "fade out" instead of disappearing in an instant.

The moon is, therefore, a dead world from which all life, if it ever existed there, departed many millions of years ago. It may be added that there is no trace of water or vapour on the moon. Without an atmosphere to screen the heat of the sun, the side of the moon facing the sun gets very hot—hotter, it has been estimated, than boiling water. On the other hand, the sunless side is very cold—considerably more than a hundred degrees below the freezing point of water.

Physical Appearance of the Moon. A small telescope or a pair of field-glasses will reveal the wonderful beauty of the moon's surface, with its rugged mountain ranges, its plains, and, most wondrous sight of all, its thousands of craters. These "craters" have very little in common with the volcanic craters on the earth, which are few and also small in comparison with those on the moon. More than 30,000 lunar craters have been mapped up to the present. Their diameters vary from about 150 miles down to less than one-fourth of a mile, and there must be many others so small as to be invisible even in our largest telescopes. The title "craters" may be unfortunate, as it conveys the idea of openings on the top of mountain cones from which molten material is occasionally ejected. The larger craters are great walled plains, their rims rising as high as 10,000 to 20,000 feet above the general surface of the moon, while their interior is far below the moon's surface—sometimes as much as three miles.

There has been much speculation about the origin of the lunar craters, but it is impossible to say definitely how they were formed. Sir George Darwin suggested a theory many years ago about the origin of the moon and also about how the craters were formed. He believed that some thousands of millions of years ago, when the earth rotated more rapidly than it does now, the moon was thrown off, and has been receding ever since. If this did occur it seems that the moon would be of similar material to the outer portions of the earth, and actually the density of the moon is nearly the same as that of the earth's outer crust (see p. 8). At the time when the moon separated from the earth it is believed that the earth was in a molten condition, and of course the moon was likewise, but as it was so much smaller than the earth, it cooled more rapidly. As we should expect this cooling to occur first near the surface, where in time a solid crust would be formed (the interior remaining molten for a longer period), it has been thought that something like enormous bubbles were formed on the cooling exterior. When these burst, craters like those which we now see on the moon would be formed.

This theory is open to a number of objections. Darwin himself admitted that it was "only a wild speculation, incapable of verification," and indeed in recent times it has been shown to be extremely doubtful whether the moon broke off from the earth, as Darwin suggested.

Another explanation of the lunar craters is that there was once volcanic activity on the moon, but on a scale so enormous that the small volcanic activity on the earth can give no conception of it. Many objections can be made against this solution of the puzzle, but it will be sufficient to point out that if things happened so, there should be some trace of the enormous amount of material thrown out by the volcanoes; in most cases such material is lacking.

Another theory, probably more feasible than the other two, is that the lunar craters were formed by bombardment by meteorites which varied in size from small bodies up to objects many miles in diameter. A large object crashing on the moon at a speed of 30 miles a second or more would have an effect like a bomb and would produce a large crater. Even if the body struck the moon on the slant, the crater produced, assuming the effect is of an explosive nature, would still be circular, as most of the lunar craters are. We have a crater formed in this way on the earth—the Arizona crater—which is about 4,000 feet in diameter and 600 feet deep, and it is certain that the meteorite which was responsible for it fell obliquely, at an angle of about 45°. The theory of the formation of the lunar craters by meteorites would be strengthened if there were many more of these large depressions on the earth, formed by the impact of meteorites. On the other hand, it is possible that if a large number of such craters had been caused in this way, all traces of them might have been destroyed by the action of air and water in the course of hundreds of millions of years. No such action could take place on the moon, where, as we have seen, there is no atmosphere or water; hence if craters had been formed on her surface by bombardment they would remain intact. On the whole, the theory of meteoric action is the most plausible explanation of the craters on the moon.

The Moon and the Tides. The most important effect which the moon has on the earth is in the production of the tides. The pull of the moon on the ocean waters tends to draw them slightly towards her. Indeed, the surface of the earth is pulled as well, but as it is too rigid to give way very much, the chief visible effect is on the water. If the moon merely heaped up the waters on the side of the earth next to her, we would have a single high tide sweeping over the seas once in every 24 hours. The action of the moon, however, heaps the waters up on the far side of the earth as well as on the near side, thus giving two high tides every 24 hours. Why it should do so is rather difficult to explain without going into abstruse mathematics,* but there is no doubt about the fact, which anyone living at the seaside or on a tidal river can verify for himself (see Fig. 4).

The sun plays a part as well as the moon in tide-making. In spite of his enormous mass compared with that of the moon, his effect on the tides is less than that of the moon. This is because he is so far away from the earth—nearly 400 times as far as the moon. When the sun and the moon pull together we get the big rise of the

Fig. 4.—The moon draws the waters of the ocean towards her and hence produces the tides. On the other side of the earth there is also a high tide at the same time.

" spring tides," and when they pull against each other we get " neap tides," in which the difference between high and low water is less. During spring tides the sun, moon, and earth are in a line, or nearly in a line, with one another, and the tides made by the sun are added on to those made by the moon at the same place. At other times, when the moon is out of line with the earth and sun, the moon and sun each makes a tide of its own at different parts of the earth, but the tide produced by the moon is the greater of the two. That which the sun causes diminishes the moon tide, and so we have low tides at a place—the neap tides. These changes in position of the sun and moon account for the variations in the tides, which can be predicted and are listed in

* I gave a fairly elementary mathematical proof of the subject in *The Journal of the British Astronomical Association*, **35**, 2, November, 1924, and also **52**, 8, September, 1942, but the dynamics of these papers are beyond the scope of this work.

Tide Tables prepared for the use of mariners and harbour authorities.

The Moon and the Weather. It is remarkable that many people still believe that there is some connection between the moon and the weather. Weather records, however, show that this view is utterly devoid of foundation. Most people who hold the belief fail to carry out any regular observations, and they are impressed with the cases where good weather follows a change in the moon, but ignore those cases where no such connection is found. Sometimes it is said that if the new moon comes in with fine weather we may expect a spell of " set fair." It is forgotten by many that we can generally see the new moon only if the weather at the time is fine, and the meteorological conditions may then be favourable for a period of good weather. The meteorological conditions, however, were not determined by the moon, but merely happened to coincide with her phase. If there is a spell of wet weather at the time of new moon, we shall not be able to see her, and this spell may continue for several days. But the continuation of the wet weather has no connection with the change in the moon.

Questions

1. What are the chief uses of the moon to the inhabitants of the earth?

2. Give reasons for believing that the moon moves round the earth.

3. Suppose that you were asked by someone utterly ignorant of astronomy to explain the phases of the moon, how would you proceed? Could you explain from a model why the horns of the moon always point *away from* the sun?

4. Why do you think that the moon has no atmosphere?

5. State briefly the theories to explain the origin of the lunar craters.

6. Does the moon affect the weather?

CHAPTER III

THE SUN

Size and Temperature of the Sun. Among the myriads of bodies which are scattered about in the universe the sun is by far the most important of all for us. Without the heat of the sun no life could exist on the earth, which would then be intensely cold, even the oceans becoming a great mass of ice. Yet the sun is merely one of millions of millions of stars and only an ordinary star, not merely in size, but also in mass and temperature. Some people find it difficult to think of the sun as a star because its size and brightness are in such contrast to the smallness and feeble light of the stars. The contrast, however, is due to the fact that the sun is so close to us compared with the distance of even the nearest star. The average distance of the sun from the earth is only 93 million miles—we say " only " because the nearest star is about 270,000 times as far away—and so, as against the distance of a star, the sun is very close to us. We know a lot about the sun, not only about his composition, his temperature, his surface features, and so on, but also about his size, density, weight, etc. It is not possible to describe how all the facts are found out, but it will be sufficient to give a summary of the chief things which we know with absolute certainty.

The sun's distance from the earth varies from 94,556,000 miles to 91,444,000 miles, and it attains these distances early in July and January respectively. His diameter is 864,000 miles and his density 1·41. In spite of this low density, his mass is enormous and is represented in tons by the figure 2 followed by 27 ciphers, which is 330,000 times the mass of the earth. The temperature of the sun at his surface is 6,000 degrees on the centigrade scale—written 6,000° C.—but, high as this appears, it is very low compared with the temperature near the sun's centre. It has been estimated that this must be about 20 million degrees Centigrade or even higher. It is difficult for our minds to grasp fully the meaning of these figures, but the following simple illustration will assist us in understanding the meaning of such a temperature. Suppose we had an ordinary-sized stove which could withstand any temperature, however high, and we raised it to 20 million degrees Centigrade. Within a radius of hundreds of miles everything would be burnt up by the intense heat.

Origin of the Sun's Heat. If the sun is continually sending out heat from his surface, which has a temperature of 6,000° C., it might seem merely a matter of time before he burns himself up completely. It is not really correct to say that the sun is " burning," and the energy which supplies the heat does not arise, as it does in a fire, from ordinary chemical action. Radium and other radio-active bodies are known to give out an enormous amount of energy when their atoms break up, and other elements are capable of doing the same under certain conditions. This *subatomic energy* accounts for the heat radiated by the sun and by the other stars as well, so we need not fear that the sun will soon burn himself out and leave our planet and the others to perish from the intense cold. We need not fear, on the other hand, that the sun will suddenly blaze forth on account of all this enormous amount of subatomic energy and burn up the earth.

Rotation of the Sun. The sun rotates on his axis just as the earth does, from west to east, but takes longer to complete a rotation. In the neighbourhood of the sun's equator it requires about 25 days to move round once, but the strange thing is that it takes a *longer time* in places north and south of the equator, and at the poles the rotation is so slow that 34 days are necessary. This fact suggests that the sun cannot rotate as a whole like the earth ; different portions of his surface are moving round at different speeds. This is the sort of thing that could happen with a body made of gases, and the sun, like all the stars, is gaseous, the high temperature preventing anything from becoming solid either in the interior or at the surface.

The Sun's Surface. The study of sunspots is a

favourite hobby with some amateur astronomers, and a large telescope is not necessary to carry out the observations. If you ever use a telescope, however small it may be, to look at the sun, make sure that you have a piece of smoked glass between your eye and the eyepiece of the telescope. If you do not take this precaution you may receive very serious injury to your eye. (It is possible to see large spots on the sun by using a piece of smoked glass without a telescope, as the glass shuts out some of the glare of the sun.) These spots are very useful for determining the times of rotation of the different parts of the sun. It has been found that sunspots run in cycles over a period of 11 years. When the sunspots are at their lowest it is possible that for months the sun's disc shows only a few or perhaps no spots. Then they appear in increasing numbers, until the greatest number is reached in about $4\frac{1}{2}$ years. The numbers decrease after this for about $6\frac{1}{2}$ years, when the spots are few or entirely absent. After this they start to increase again, and pass through the same changes in the same period. There is still much doubt about the causes of these changes.

Sunspots are much cooler than the rest of the sun's surface, and they look dark only by contrast with the brighter regions which surround them. They are caused by whirling columns of gases which rise from beneath the sun's surface and lose heat by their expansion. At the same time warmer gases from the sun's surface are drawn into the whirlpools produced by the rising gases, and a sort of circular motion takes place. When the sunspots are very active the magnetic compass is affected and there are displays of the aurora or " northern lights," while radio service is often upset. There is no mysterious connection between the sunspots and " magnetic storms " as they are often called. They are due to the arrival of showers of electrons or small negative charges of electricity from the disturbed part of the sun. When these electrons reach our atmosphere about two days after leaving the sun they produce certain electrical conditions which disturb our electrical and magnetic instruments. Sometimes the disturbances upset our telegraphic and telephonic services, but they are never dangerous to life.

Eclipses. The model described in the previous chapter will help to explain how and when eclipses of the sun and moon occur. With the moon revolving round the earth, and the earth and moon round the sun, there are certain positions in which the earth's shadow falls on the moon (giving an eclipse of the moon), and other positions in which the moon comes directly between the earth and the sun (giving an eclipse of the sun). If the moon moved all the time in the plane of the ecliptic there would be an eclipse of the moon every full moon and an eclipse of the sun every new moon. We know, however, that the path of the moon is inclined to the plane of the ecliptic. Owing to this tilt in the moon's orbit, the three bodies— the sun, moon, and earth—do not come into line regularly at new moon and full moon, but at less regular intervals. Astronomers are able to calculate the times when these events occur, and so are able to predict the occurrences of eclipses hundreds of years before they take place.

Here are two rules worth remembering in connection with eclipses :—

> (1) In one year there cannot be more than seven eclipses: five of the sun and two of the moon, or four of the sun and three of the moon.
>
> (2) In any year there cannot be less than two eclipses of the sun.

Eclipses are not always total—that is, only part of the sun may be hidden by the moon in a solar eclipse, and only part of the moon may be covered by the earth's shadow in a lunar eclipse. We then speak of the eclipses as " partial." A total eclipse of the sun is a great event for astronomers. It is a fortunate circumstance that the moon is of such a size and at such a distance from the earth that it appears to be practically of the same size in the heavens as the sun. At times it just covers the sun's disc, and astronomers are able to obtain a better view of certain portions of the sun's atmosphere that extend far beyond the disc, and in particular of the corona—a pearly white envelope of wonderful beauty which can be seen best during a total eclipse. By photography and other means much valuable information is obtained about the atmosphere of the sun which cannot be obtained at other times.

During solar eclipses other than total the moon is a little farther away from the earth, and some portion of the sun's disc is visible round the moon. Such an eclipse is called a partial eclipse. It has not the same value to the astronomer as a total eclipse.

As a total eclipse of the sun is such an important event, astronomers make expeditions to different parts of the earth where it can be seen. A total eclipse of the sun is not visible everywhere, but over only a very limited portion of the earth's surface. The width of the shadow thrown by the moon on the earth is only about 50 miles, and anyone inside this strip, which sweeps over a great part of the earth's surface, can see the eclipse as total. Those who are just outside the strip can see it only as a partial eclipse, and hence it is important for astronomers who want to see a total eclipse to know exactly where this strip runs. Not every astronomer is able to do the calculation necessary to determine the precise track and give the instant when the eclipse begins or ends. This is very specialized work for the mathematician, and so great is the accuracy of his computations that he can predict to within a second or two the time of beginning and ending of an eclipse at any place. Mistakes would be very disturbing to those who go to the other end of the earth with elaborate appliances to see and photograph a total eclipse of the sun (a lunar eclipse has not the same value as a solar eclipse), and woe to the mathematician who made even the slightest error! Such a thing, however, never occurs, and when an eclipse is predicted for a certain time and place you can be absolutely sure that it will take place then and there.* The one thing which the mathematician or anyone else cannot predict months ahead is the condition of the weather. Unfortunately even a passing cloud will prevent astrono-

* The next total eclipse of the sun visible in the British Isles (actually only in Cornwall) will be on August 11, 1999.

mers from seeing an eclipse and will spoil all their
months of preparation, but this will not deter them from
going off on future expeditions when other eclipses are
due. (See Figs. 5 and 6.)

Gravity of the Sun. In the chapter on the moon we

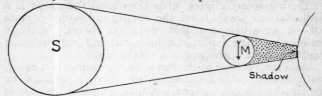

FIG. 5.—Showing a total eclipse of the sun. The moon casts a
shadow on the surface of the earth, and this shadow sweeps
rapidly over the earth as the moon revolves in the direction
shown by the arrow.

spoke about the force of gravity there—about one-sixth
of what it is on the earth. If you were on the sun you
would find conditions just the opposite. The force of
gravity on the sun is 28 times as much as it is on the earth,
so if your muscles were just as strong there as they are

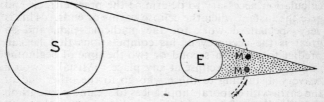

FIG. 6.—An eclipse of the moon. The earth casts a shadow and
the moon is shown in two positions inside this shadow. An
eclipse of the moon can occur only at full moon.

here you would find it difficult to move about. Instead
of jumping over a high wall, as you could do on the
moon, you would be lucky if you could jump 3 inches
from the sun's surface. A man who weighed 11 stone on
the earth would weigh nearly 2 tons on the sun, *provided
a spring balance specially designed for heavy weights was
used.* If an ordinary weighing machine was used, the
man would be balanced with 11 stone, because, although

he weighs 28 times as much as this, the same is true of the 11 stone.

Time Required for Light to Reach Us. The sun is never exactly where we see it, because the light takes an appreciable time to travel to the earth. The speed of light is 186,000 miles each second, and in the case of the moon less than $1\frac{1}{2}$ seconds is required for the light to reach us, but in the case of the sun the time is about 8 minutes. The nearest star is so far away from us that the light from it takes about $4\frac{1}{3}$ years to reach the earth. When we look at the sun at any time we see it where it was 8 minutes ago. In certain astronomical computations—for example, in finding the path in which a comet moves round the sun—it is necessary to make corrections for the "light-time," and before doing so the distance of the comet from the earth must be determined. When this has been done we are then able to make the necessary corrections and to find the path of the comet accurately. Without such corrections small errors would be introduced into the work. Such are some of the refinements in astronomical investigations.

Questions

1. Give the following established facts with reference to the sun: (*a*) its distances at the beginning of January and July, (*b*) its mean distance, (*c*) its diameter, (*e*) its mass, (*f*) the temperature of its surface.

2. From surface markings, such as sunspots, it is known that the sun's rate of rotation is not the same all over its surface. What conclusion would this suggest to you?

3. What is the cause of sunspots? State briefly what effect they have on the earth.

4. What is the greatest number of eclipses of both the sun and the moon which can occur in a year?

5. On August 1, 1943, the moon was in the plane of the ecliptic, and new moon occurred on the same day. What interesting phenomenon took place?

6. On August 15, 1943, the moon was again in the plane of the ecliptic and full moon occurred on the same day. What happened?

7. Find to the nearest second the time that light takes to travel from the sun to the earth on January 2 and July 4, 1944, the dates when we are nearest to and at the greatest distance from the sun.

8. Compare the results in 7 with the times which an aeroplane flying at the rate of 3 miles a minute would require.

9. If the distance to the nearest star is 4·3 light-years, how many miles is the star away? (A light-year is the distance travelled by light in one year.)

CHAPTER IV

THE SUN'S FAMILY OF PLANETS

MERCURY

Mercury Has no Atmosphere. The earth is only one out of a family of nine planets which belong to our solar system—that is, which move in orbits round the sun in a similar manner to the earth. Besides the nine chief planets there are thousands of very small planets, known as the asteroids or minor planets, but few of these can be seen without the help of a fairly powerful telescope. They will be considered in the next chapter.

The nearest planet to the sun is Mercury, and it is also the smallest of the nine planets. Its diameter is only 3,000 miles, which is half as much again as that of the moon. We saw that the moon had not got an atmosphere, and the same thing is almost certainly true of Mercury. On the moon the velocity of escape is 1·5 miles a second, and on Mercury it is a little more—2·4 miles a second—so if there were an atmosphere on Mercury it seems that it would have more chance of being held by the planet's attraction than in the case of the moon. But the temperature of Mercury is much higher than that of the moon, and for this reason the gases composing the atmosphere would be more likely to escape, because the speed of the molecules increases as the temperature increases. We may take it as almost certain that Mercury has not got an atmosphere, and hence that no life exists there.

Elliptical Orbit of Mercury. We saw that the earth is not always at the same distance from the sun ; the same is true of Mercury. In fact there is a much bigger difference between the greatest and least distances of Mercury than there is with the earth. Mercury comes to a distance of 28½ million miles from the sun at one time and goes out to a distance of over 43 million miles 44 days later (the planet takes 88 days to make a revolu-

tion round the sun), and as a result of the large difference
in its distances from the sun there are great changes in its
surface temperature.   Unlike the earth, which rotates
on its axis in 24 hours, and so enjoys the succession of
day and night everywhere, the planet Mercury always
turns the same face to the sun.   If there were people on
Mercury, those on one side would never enjoy sunshine,
and those on the other side would never lose sight of the
sun.   It would be very trying to live there in perpetual
light, but worse still to live in perpetual darkness.   This
would not be the only inconvenience.   On the sunny side,
if you were nearly under the sun, as people are on the
earth near the equator, the temperature would be nearly
280° C. when the planet was at its greatest distance from
the sun, but when it made its nearest approach the
temperature would be as high as 400° C. or even higher.
Meanwhile people on the dark side would suffer the most
intense cold, far beyond anything we have ever endured
at the coldest place on the earth.   Perhaps it is just as well
that there is no life on Mercury.   Conditions are sometimes hard enough on our own planet, but they would be
infinitely worse on Mercury.

Mercury Invisible by Night.   You can see some of the
planets at midnight, or indeed at any time of the night,
but you will never be able to see Mercury at such hours.
The planet is never far away from the direction in which
we see the sun, and hence is lost in the glare of the sun's
light unless a telescope is used.   If you knew exactly
where to set your telescope during the day you would be
able to pick the planet out, in spite of the light of the sun;
but this is very difficult, and we do not recommend
amateurs to search for Mercury, especially as there is not
very much detail that can be observed with a small
telescope.

VENUS

Resemblances between Venus and the Earth.   The next
planet as we go out from the sun is Venus, which is very
like the earth from the point of view of size and mass.
Venus does not behave like Mercury in approaching the
sun much closer at one time than at another; in fact the

orbit is nearly circular. The greatest and the least distances of Venus from the sun are 67½ and 66½ million miles, and the time required to complete a revolution is 224 days. The diameter is 7,600 miles and the mean density 5·21. So you will see, by comparing these figures with those given for the earth, that there is a close similarity between the two planets. Even in their masses this likeness holds, as you would infer from their diameters and densities. The mass of Venus is about four-fifths that of the earth, and gravity is nine-tenths of gravity on the earth. The velocity of escape is 6·5 miles a second—just a little less than on the earth—so it might be expected that Venus would be able to hold an atmosphere.

Observation shows that the planet has an atmosphere, but not of the same composition as our atmosphere. It is remarkable that, so far as we know at present, there is neither oxygen nor water-vapour on Venus, but there is carbon dioxide in abundance. No definite decision has yet been reached about the period of axial rotation. Some think it is 224 days—the same as the time of revolution round the sun—and others believe that it is about the same as our earth's period. It may seem remarkable to some that astronomers cannot fix this period more definitely, but there are good reasons for the uncertainty.

Venus Surrounded by a Permanent Cloud. If you ever get the chance of looking at Venus through a telescope, do not lose it. You may see the planet when it is a crescent like the moon, and this is a beautiful sight. You may see her when she is more like a full moon, and probably you will be disappointed with the view. The reason why the planet appears as a crescent and goes through phases something like the moon is because, like the moon, Venus sometimes comes between the sun and the earth, and in such cases we see only a slice of the illuminated surface. At other times she is, like Mercury, on the side of the sun remote from the earth, owing to the journey of the planet round the sun ; the illuminated portion is then turned towards us, and we see her like a full moon. The disappointing thing about Venus is that it is quite impossible to see the surface of the planet because of the layer of cloud which always surrounds her. When the cloud clears a little, as it does very occasionally, the eye can detect nothing on the *surface*; only badly defined shadings are seen, and these are almost certainly in the atmosphere. Of course, if there were clear markings and regular observations were carried out, it might be possible to determine the time of rotation, but up to the present this has not been accomplished with any degree of certainty.

Venus is a morning or evening star and, like Mercury, cannot be seen all the night. Under the most favourable conditions she can set in our island about four and a half hours after the sun. She is an evening star when she sets after the sun, and a morning star when she sets before the sun.

Vegetable Life Possible on Venus. The temperature on Venus is not inconsistent with the existence of life there. On the sunlit face the temperature is between 50° and 60° C., and on the other side it is about −20° C. The last temperature is lower than we are generally accustomed to on our earth during the night, but life is very adaptable, and it could certainly survive under such conditions. What the nature of the life may be, if there is any at all, it is difficult to say, and we can only speculate on the subject. The figures for the temperatures refer to the atmosphere itself, because it is impossible to measure the temperature beneath it; but, just as a greenhouse traps the heat of the sun and becomes warmer, so the surface of Venus has probably a higher temperature than is measured in her atmosphere. If the atmosphere is moist and warm near her surface there may be some form of vegetable life, but animal life, such as we know it, is not very likely. As so many of the factors on which life depends are unknown to us, much of this is pure guesswork.

THE EARTH

Proceeding outwards from the sun, the next planet is the earth. As we have already dealt with it in the opening chapter, nothing more will be added, and we shall pass on to Mars, which is the next planet beyond the earth.

MARS

Temperature of Mars. The planet Mars is the most interesting of all the sun's family in one respect—it is probably the abode of life, and this life may possibly be of an intelligent form.   Mars is smaller than the earth and its density is less than the earth's, so it is not such a massive planet as our own.   It moves in an orbit round the sun at distances which vary from nearly 155 to 128 million miles.   We have already explained that the earth's distances from the sun vary from about $94\frac{1}{2}$ to $91\frac{1}{2}$ million miles, and hence, when the earth and Mars are on the same side of the sun, the planet can be as close to us as 34 million miles or as far away as 63 million miles. When the two planets are on different sides of the sun the distance can be as great as 250 million miles.

The diameter of Mars is 4,200 miles, the density 3·9, and its mass just over one-tenth that of the earth.   Its surface gravity is 0·38 of the earth's gravity and the velocity of escape is 3·2 miles a second.   This is perilously near the " danger line " for ordinary gases which compose the atmospheres of the smaller planets, and perhaps we should not be surprised if Mars had no atmosphere.   Actually it has an atmosphere, but its amount is very much less than we have on the earth, so life there, assuming that there is such, would probably take on forms rather different from those to which we have been accustomed.

The planet moves round the sun in 687 days and rotates on its axis in 24 hours 37 minutes, so it enjoys alternation of day and night almost in the same interval as we do.   We should expect to find that the temperature on Mars was lower than that on the earth, owing to its greater distance from the sun, and in fact its temperature is much lower.   Owing to the scanty atmosphere and the consequent absence of the blanketing effect in trapping the sun's heat like a greenhouse, the nights on Mars must be very cold.   Near equatorial regions it has been found that a temperature far above the freezing point of water— perhaps as high as 30° C.—exists, but at sunrise it is as low as −10° C., and during the night it is supposed to fall to −90° C.   The temperature at the polar caps has been measured and found to be as low as −70° C., but as this refers to the cloud layer above the cap, it is possible that the actual temperature of the cap itself may be higher. We are accustomed to think of the polar caps as immensely thick, and indeed so they are on the earth, but on Mars they are only a few feet thick or even less. This is an important matter in view of the fact that some have believed canals were constructed on the planet to irrigate the arid parts by conveying the water from the poles when the caps melted during the summer on each hemisphere. We shall return to this point later in the chapter.

Atmosphere on Mars. If you ever look at Mars through a telescope you will notice that it has very little resemblance to Venus. Some parts of the planet appear reddish-yellow and others are bluish-green, and the latter are suggestive of seas—a title which the early observers bestowed on these areas. This view is no longer held, and it is thought that they may be some form of vegetation. The general ruddy colour of the planet has been attributed to the rocks, which have taken most of the oxygen out of the atmosphere, the oxidized rocks appearing red. In contrast with these the rocks on the moon look grey or brownish because they were unable to combine with oxygen, the moon being devoid of an atmosphere. Support is lent to this view by the fact that the amount of oxygen in the atmosphere of Mars is very small, but it may have had an abundance of this gas before the rocks absorbed it. How far the present supply is capable of supporting animal life is a question which it is impossible to decide. We are disposed to judge the conditions necessary for life from our own limited point of view, and hence assume that animal life would be impossible without an amount of oxygen comparable with the quantity existing in our atmosphere. This may be so, but there is no proof that it is, and there seems no fundamental obstacle to animal life—perhaps in a more intelligent form than we find on the earth—existing on Mars. There is almost certainly vegetable life there.

Surface Markings on Mars. We have already spoken about the " canals " on Mars, which some have held to be made with the object of carrying water to the dry regions of the planet. It would require too much space to discuss this matter fully, but it may be briefly summarized by saying that the notion of artificial canals, so strongly held by the late Professor Lowell and others, is now discredited. They are surface markings which the eye often interprets as continuous lines, although continuity does not always exist, and experiments have shown how easy it is for the eye to misjudge certain details.

Here is an experiment which you can try for yourself. Draw a number of dots on a piece of paper, say about five to the inch, and make them large enough to be seen about 30 feet away. Now ask someone who does not know that they are dots to stand about 30 feet from them and to tell you what he sees. You will find that he will probably say he sees a line, but if you show him the dots first of all he will almost certainly interpret them as dots, even at a distance. For the same reason, markings on Mars which appear continuous are not necessarily continuous, and it is practically certain that some eminent astronomers have been misled by appearances. Of course, the absence of the so-called canals does not necessarily imply that intelligent life is non-existent on Mars. Perhaps if it does exist there it does not find equatorial regions so arid as some have thought they are —at least not too arid for life which has accommodated itself to its environment.

Satellites of Mars. Mars has two moons or " satellites," which is the usual term for these bodies. They are very small, the greater of the two, Phobos, having a diameter about 10 miles. It revolves round Mars at a mean distance of 5,828 miles in 7 hours 39 minutes. The smaller satellite, Deimos, has a diameter about half that of Phobos, and revolves round Mars in 30 hours 18 minutes at a mean distance of 12,475 miles. These satellites, like the majority of those in the solar system, revolve in the same direction in which the planet is rotating. A curious effect is produced in the case of Phobos. Its period of revolution is about one-third the period of rotation of Mars, and hence this satellite, as seen from the surface of the planet, rises in the west, moves eastward, and sets in the east. Deimos does not act in the same way, because its period of revolution exceeds the period of rotation of Mars; and it rises in the east and sets in the west, just like our own satellite, the moon.

JUPITER

Dimensions and Atmosphere of Jupiter. Outside the orbit of Mars there are the asteroids to which we have referred in the early part of the chapter. As they are very small we shall pass them by for the present and look at the next planet out from the sun, the largest of them all, the planet Jupiter.

Jupiter revolves round the sun in a period a little under 12 years, and his distances from the sun vary from over 506 to 460 million miles, the mean distance being 483 million miles. The density is only 1·34, so that the planet weighs only one-third more than an equal volume of water. Jupiter rotates on his axis in 9 hours 55 minutes—a very short time for such a large body—and as a result of this rapid rotation there is considerable flattening at the poles and a bulge at the equator, which can be easily detected with a small telescope. The diameter of Jupiter at his equator is 88,700 miles and his polar diameter is 83,800 miles; these figures show a considerable flattening at the poles. We have seen that the difference between these two diameters in the case of our earth is 26 miles, but with Jupiter it is nearly 5,000 miles. In spite of the fact that the planet is over 1,300 times the size of the earth, its weight is only 318 times that of the earth, the apparent discrepancy arising from the low density of Jupiter. It is believed that the height of the atmosphere is a very great fraction of the planet's radius, and this view explains the disparity in the densities of the earth and Jupiter. Some authorities have thought that Jupiter's atmosphere is about 6,000 miles thick, but there is considerable doubt if it is actually as thick as this. It is also believed that outside the central rocky core (which has a diameter about half that of the planet) there is a layer of ice thousands of miles thick. It is certain that there is no evidence of internal heat in

Jupiter, and the greater part of the radiation which he sends out is only reflected sunlight. The temperature of the atmosphere is about —130° C.—which is not very suitable for life in higher forms such at least as we know on our planet. Even if the temperature was congenial to life the atmosphere is very unlike anything of which we have experience. It consists mainly of ammonia and methane (or marsh gas), the latter occurring on the earth in marshy places by the decomposition of vegetable matter. If you disturb the bottom of a pond the bubbles of gas which rise to the surface consist chiefly of marsh gas. As you probably know, it is also found in large quantities in coal mines.

The markings on Jupiter are not so permanent as those on Mars, and we should expect this because what we see is mainly atmosphere, not the solid crust of the planet. Just as the sun has different rates of rotation at different latitudes, so Jupiter shows the same characteristics. We have already spoken of his period of axial rotation as 9 hours 55 minutes, but in the equatorial zone it is 5 minutes less than this, and so any particular features close to the equator gain on those which are north or south of this line.

Surface Features of Jupiter. A small telescope shows a few dark streaks—the *belts*—which are parallel to Jupiter's equator. When a large telescope is used these belts exhibit a very large variety of detail and rapid changes of form. Of course such changes would not occur on a solid surface, and what is seen is something moving in the planet's upper atmosphere. The most conspicuous of the belts are the *north and south equatorial belts*, which are about 10° on either side of the equator and include the bright *equatorial zone* within their boundaries. Beyond the equatorial belts are the bright *north and south tropical zones*, and beyond these there are less conspicuous belts and zones until we come to the polar regions. The study of the Jovian features has been the life work of some astronomers who have specialized on this subject, but in spite of all their work there is still much that we do not know about the planet, but which future observers will no doubt be able to explain.

Satellites of Jupiter.   When you look at Jupiter with a small telescope you will be struck with the beauty of the satellite system.   Four out of the eleven satellites which attend the planet are visible in the smallest telescope; they were discovered by Galileo in 1610 with his home-made telescope.   They move in nearly circular orbits round Jupiter and also very close to the plane of his equator, which is near the plane of the ecliptic.   When you see them through a small telescope, or even through a pair of good binoculars, they appear just like faint stars close to the planet.   They may all be on one side or they may be on different sides of Jupiter, because they all have different periods of revolution round the planet, and so are seen in various positions with regard to their parent.   These satellites move with great regularity, and it is possible to predict exactly where any one of them will be at a certain time.   Just as eclipses of the sun and moon can be predicted, so the eclipses of Jupiter's satellites can be predicted too.   These eclipses occur when a satellite moves into the shadow thrown by Jupiter in the same way as the moon is eclipsed when she moves into the shadow thrown by the earth, the sun's light being intercepted in each case.

In 1675 the Danish astronomer Roemer noticed that the observed times of eclipses occurred earlier than the predicted times when the earth was nearer Jupiter than the average, and also that the reverse occurred if the earth was at a greater distance from Jupiter than the average. Before his time it was known that predicted and observed times did not agree, but no adequate explanation was forthcoming until he pointed out that light did not travel with an infinite velocity, as was believed prior to his days. When the earth was nearest to Jupiter, light would not have so far to travel, and so the eclipse would appear earlier than was expected.   On the other hand, when the earth and Jupiter were as far apart as possible, the light would have a greater distance to travel and the eclipse would be seen after their predicted times.   This discovery afforded one method of measuring the speed of light. (See Fig. 7).

The name of the four largest satellites of Jupiter are

Io, Europa, Ganymede, and Callisto. The last two are
the largest, having diameters over 3,000 miles, which is

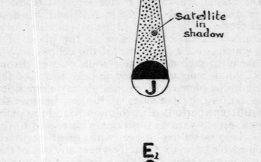

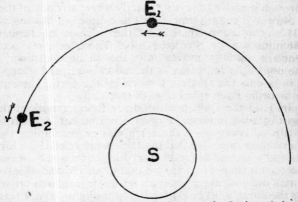

Fig. 7.—The diagram shows the shadow cast by Jupiter, and one
of his satellites in the shadow. When the earth is in position
E_1 the eclipse of the satellite will be observed sooner than
when the earth is in position E_2. This is due to the fact that
the distance from E_1 to Jupiter's satellite is less than that from
E_2 to the satellite, and hence the light takes longer to travel to
the earth in the latter case.

half as large again as the diameter of our satellite.
Eight of Jupiter's family of satellites have "direct"
motion—that is, they move round their parent in the
same direction in which he, in turn, moves round the

sun. Three have "retrograde" motion; they move round Jupiter in the opposite direction to which he moves round the sun. Very important problems are raised by retrograde motion, which, as we shall see, occurs in some other cases as well, but it is beyond the scope of this work to discuss these.

SATURN

Saturn the Least Dense of all Planets. Saturn is the next planet as we journey out from the sun, and it ranks close to Jupiter in size, though not in mass. Its equatorial and polar diameters are 75,000 and 67,000 miles respectively, and its rotation period is 10 hours 14 minutes. The rapid speed of rotation accounts for the differences in the two diameters and, as with Jupiter, the polar compression or flattening can be clearly seen through a small telescope.

Saturn's greatest and least distances from the sun are 931 and 841 million miles respectively, the mean being 886 million miles. To complete a journey round the sun requires nearly 29½ years. It has the least density of all the planets, being only seven-tenths the density of water. It is startling to realize that if Saturn could be placed in an enormous ocean it would float; it is the only planet that could do so. As its density is so low it follows that most of what we see of Saturn is simply its atmosphere. It is believed that it resembles Jupiter in this respect, and that, underneath an extensive atmosphere—which, like Jupiter's atmosphere, is composed of ammonia and methane—there is a thick layer of ice, and underneath this a rocky core. The clouds are arranged in bands, as they are on Jupiter, but the bands are much less distinct (which is not surprising in view of the planet's greater distance), and they do not appear to change their shape so rapidly. Underneath the envelope of clouds the planet itself is wrapped in mystery, and we can only guess about its physical condition.

If Saturn itself does not present many attractive features, the same cannot be said about its system of rings, which we shall now describe.

Saturn's Rings. There is no more beautiful sight in all the heavens than the rings of Saturn, and no one should

D

miss an opportunity of looking at them through a telescope. A small telescope of about 3 inches aperture will show them, provided they are well placed for observation. The ring system is 171,000 miles across and is made up of a dusky ring, an inner light ring, and an outer light ring.

About 7,000 miles from the equator of Saturn the crape or dusky ring begins, and it extends for 11,500 miles, after which there is a gap of 1,000 miles between it and the inner bright ring. The inner bright ring is 16,000 miles across; beyond it there is another gap (Cassini's Division) under 3,000 miles wide, and then we come to the outer bright ring, the width of which is 10,000 miles. It will be seen from these figures that the total width is 41,500 miles, but if to this is added 7,000 miles—the distance of the crape ring from the surface of the planet—and then to this 37,500 (the radius of Saturn), we find that the far edge of the outer bright ring is about 86,000 miles from the centre of the planet. Hence, if we measure across from one extremity of the rings to the other, the distance is about 171,000 miles, though of course only 82,000 miles of this consists of the rings. The thickness of the rings is probably about 10 miles, but it is impossible to be very definite about this, and it may be more than 10 miles, but is almost certainly less than 50 miles.

When the rings are so situated that their broadside is partly towards us we see them better than when they are turned edge on. In fact, in the latter case they disappear almost completely and look like a narrow bright line. You can illustrate this by taking a piece of cardboard on which light is shining and turning its broadside towards you. It will be seen easily at some distance but will not be so visible if the edge is turned towards the observer.

Nature of the Rings. When we speak of Saturn's rings the term is slightly misleading because we generally think of a ring as something solid. But Saturn's rings are not solid; they are composed of myriads of small bodies, some as small as dust particles and some probably the size of a cricket ball, but none of them large. How did they come to be where they are, every small body pursuing its own orbit round the planet, just as the planets pursue their own orbits round the sun? There is no reason why they should not continue to follow their present orbits for thousands of millions of years, but there was a time when conditions were very different. It is fairly certain that when we look at this beautiful ring system we see the fragments of a former satellite whose eagerness to cultivate too close an acquaintance with the planet was its own undoing. Once it was moving round Saturn as the moon moves round the earth or as Saturn's other satellites still describe their orbits round the planet. It came, however, too close to the planet, with disastrous results. There is a law in celestial mechanics that it is dangerous for bodies to come too close together. Like relatives who live apart, peace may be preserved so long as plenty of space intervenes, but if the space narrows, trouble may arise. So with celestial bodies; if they venture too close together the smaller will be torn into pieces by the larger and more massive body. At some time a small satellite must have gone into the danger zone near Saturn, and we see the results of this step in the fragments composing the rings. The sight of the rings is so wonderful that no astronomers have ever been heard to express any regrets at the fate of the little satellite whose expedition too close to its giant companion ended in its destruction.

Saturn's Satellites. Of Saturn's nine satellites the sixth in order from the planet is Titan, which is the largest of them all. Its diameter is 3,550 miles and its mean distance from Saturn is a little over 759,000 miles. It requires 16 days to move round Saturn. The next in size is Rhea, the fifth in order from Saturn, with a diameter of 1,150 miles. The diameters of the others are all under 1,000 miles. The satellites of Saturn revolve in direct orbits, except the outermost one—Phoebe—which is over 8 million miles away from its planet and completes a revolution in 550 days.

URANUS

Uranus was discovered accidentally in 1781 by Sir William Herschel when he was observing a region of the sky in the constellation of Gemini with his 7-inch reflecting telescope. At first he thought it was a comet. In those days no one expected to discover a new planet, and several months passed before astronomers were sure that a new world outside the orbit of Saturn had been found.

Uranus has a year which is 84 of our years; that is, the planet takes 84 years to complete a circuit round the sun. The average distance of the planet from the sun is 1,783 millions of miles, which is 19 times the distance of the earth. Its diameter is 31,000 miles, and it rotates on its axis once in 10 hours 48 minutes. Its density is 1·36— just a little greater than that of Jupiter. The atmosphere consists chiefly of methane; large telescopes show belt-like markings which are probably of the same nature as are those of Jupiter and Saturn. There are four satellites attending Uranus, and their revolution as well as the rotation of the planet has certain interesting features. Generally speaking, satellites revolve in orbits which are not very far from the plane of their planet's equator. The plane of a planet's equator is usually not inclined at a great angle to the plane of the ecliptic. In the case of the earth this angle is $23\frac{1}{2}°$. For Mars it is nearly the same, while it is very small for Jupiter. The plane of the equator of Uranus is, however, nearly at right angles to the plane of the ecliptic, and as the satellites revolve close to the plane of the planet's equator, their orbits are nearly perpendicular to the ecliptic.

You will be fortunate if you see Uranus with the naked eye, as it is just on the borderline for vision without optical aid. Unless it is viewed through a large telescope no details of its surface features can be seen. Even through a large telescope its satellites appear so small that it is impossible to measure their diameters.

NEPTUNE

Neptune revolves round the sun once in 165 years at a mean distance of 2,793 millions of miles. Its day is 15 hours 40 minutes in length, but it is practically certain that there is no form of life on the planet to enjoy day and night. Its temperature has not been measured directly, but as it is much farther from the sun than Uranus, the temperature of which has been found to be −180° C., we may be certain that the planet is colder than this and, like Uranus, incapable of sustaining life. Its atmosphere is similar to that of Uranus, consisting largely of methane.

The diameter of Neptune is 33,000 miles and its density is 1·32. It is attended by one satellite, named Triton, which revolves with retrograde motion. Very little is known about the surface features of the planet. It is invisible to the naked eye.

PLUTO

Pluto was discovered on March 13, 1930, not accidentally but as a result of a long and systematic search for a planet beyond Neptune. Mr. Clyde Tombaugh, at Lowell Observatory, while examining photographs taken in January of 1930, detected the planet on the plate.

It revolves in an orbit at a mean distance of 3,670 million miles from the sun, completing one revolution in 248 years. This orbit is very eccentric and the distances from the sun vary from 4,587 to 2,752 million miles. Comparing this last distance with that of Neptune it will be seen that at times Pluto is actually closer to the sun than Neptune, though at other times it is half as far again from the sun as Neptune. Nothing is known about its period of axial rotation, and its satellites, if it has any, would be very much too small to be visible. It is difficult to measure its diameter, but the planet is probably about the size of Mars.

Questions

1. Why do you think that Mercury has not got an atmosphere?
2. Explain why great contrasts occur in the temperature of Mercury.
3. Why is it impossible to see Mercury a long time after sunset?
4. What is the chief gas in the atmosphere of Venus? What kind of life would you expect with such an atmosphere?
5. Explain why Venus sometimes appears as a crescent and sometimes like the full moon.
6. Why would you expect the atmosphere on Mars to be very thin?

7. What are the "canals" on Mars?

8. Is there any explanation of the ruddy appearance of Mars?

9. What interesting facts do you know about Phobos, and what phenomenon would people on Mars see in connection with this satellite which we do not see in the case of the moon?

10. Explain the polar flattening and equatorial bulge on Jupiter.

11. What would you infer about the atmosphere of Jupiter from the fact that the density of the planet is so small?

12. Why would it be impossible for life, such as we know it, to exist on Jupiter?

13. How have Jupiter's satellites been used to measure the speed of light?

14. What is there unique about Saturn?

15. Give a description of Saturn's rings. Of what are they composed?

16. Why are the rings sometimes invisible?

17. What is the past history of the rings?

18. Who discovered Uranus, and how was he mistaken about the planet?

19. What is there peculiar about the satellites of Uranus?

20. What do you think about the possibilities of life on Neptune?

21. When was Pluto discovered? What is its mean distance from the sun?

CHAPTER V

THE ASTEROIDS, COMETS, AND METEORS

Discovery of the Asteroids. Between the orbits of Mars and Jupiter there lie an immense number of small bodies of various sizes, all moving in different orbits at various distances from the sun. These bodies are known as the *asteroids,* or *minor planets* or *planetoids.* Only one of them—Vesta—can be seen with the naked eye, and as far as appearances go these bodies are of little interest. Nevertheless they present problems of great importance to astronomers, and so something must be said about them.

The first to be discovered is called Ceres, and its discovery by Piazzi, in Sicily, on January 1, 1801, was quite accidental. Pallas was discovered in the following year by Olbers. Two more, Juno and Vesta, were discovered in 1804 and 1807 respectively, and then a long interval passed before any others were detected. An amateur astronomer, Hencke, searched for 15 years in the hope that he would be able to find more minor planets, and in 1845 his efforts were rewarded by the discovery of the fifth of these small bodies. Every year since 1847 has added to the list, and at present there are over 2,000 of them known. Search for them is now conducted by photography. There must be scores of thousands of them which have not yet been found, but probably the larger ones have all been discovered.

The diameter of Ceres is 480 miles, and Pallas, the next in size, has a diameter of 300 miles. Most of these bodies are less than 50 miles in diameter and the majority are very small—some of them only a few miles across. They do not all move close to the plane of the ecliptic like the other planets, and their orbits are often very flat ovals. One of them, Hidalgo, is nearly five times as far from the sun at the greatest as at the least distance, and the inclination of the orbit to the plane of the ecliptic is 43°. A few of the asteroids have come very close to the earth, but there is only a very remote possibility of a collision ever occurring.

Origin of the Asteroids Unknown. There is a certain similarity between the asteroids and Saturn's rings, though the asteroids are greatly outnumbered by the bodies making up the rings. The similarity does not mean that the asteroids were formed in the same way as the rings. In fact the asteroids are not close enough to Jupiter—their nearest big neighbour—for disruption to have been caused by his attraction. Nevertheless, it is fairly certain that the asteroids were caused by some celestial catastrophe, but the nature of the catastrophe is doubtful. Professor K. Hirayama of Tokio showed that the asteroids could be divided into five families so far as their orbits were concerned, and that all the asteroids in each family probably came from the same source. This would imply that the explosion of a single planet could not have been responsible for the production. Perhaps there was once a single planet which divided before it had become solid during its cooling stage; and then each of these parts suffered from some internal convulsion which broke them to pieces. This is merely guesswork, and the astronomer still finds the problem of the origin of these bodies very puzzling.

Bode's Law. Johann Elert Bode, a celebrated German astronomer, was born in 1747. From his earliest days he was devoted to the mathematical sciences—astronomy in particular. His law * on the distances of the planets from the sun is useful in helping us to remember these distances, taking the earth's distance from the sun as the " yardstick."

Write down the numbers 0, 3, 6, 12, 24, and so on, each number after the second being double the one before it. Add 4 to each of these and we get a series of figures—4, 7, 10, 16, and so on—which will give the relative distances of the planets from the sun, except in the cases of the two outer planets, Neptune and Pluto. The figures are not exact, but they are fairly close. Writing down the numbers according to the directions it will be

* Titius is associated with Bode in formulating the law.

seen from the following table that they correspond closely in many cases to the mean distances of the planets.

Mercury	Venus	Earth	Mars		Jupiter	Saturn	Uranus	Neptune	Pluto	
4	7	10	16	28	52	100	196	388	772	(According to Bode's Law)
3·9	7·2	10	15·2		52	95·4	192	300·7	390	(Actual Distances)

If 10 represents the distance of the earth from the sun, the distance of Mercury should be 4, of Jupiter 52, and so on, so that Jupiter should be 5·2 times as far away from the sun as the earth is, always remembering that *mean distances* are used. The figures in the lower line show the *actual distances*, that of the earth being 10, and they are in good agreement with those computed by Bode's Law, except for the last two planets. When the law was first put forward nothing was known about the asteroids or the three outer planets, and when Uranus was discovered it was found to fit in very well with the scheme. It was noticed that there was a gap at 28, and this convinced many astronomers that there must be a missing planet at that distance, between the orbits of Mars and Jupiter. A society of astronomers was formed to search the heavens systematically for the missing planet. When the asteroid Ceres was discovered it was found to fit in with the distance; the average asteroid fulfills the conditions for the distance 28 fairly well. (See Fig. 8.)

No satisfactory explanation has been given of Bode's Law, but it cannot be considered a mere coincidence in spite of the fact that it does not apply to Neptune and Pluto. Perhaps when we know more about the origin of the planets it will be possible to explain the law.

COMETS

Comets Once Portents of Disaster. The appearance of comets in ancient times was usually regarded as an omen of some disaster, and historians were careful to record their appearances. Even in more recent days there have been cases of superstitious dread of these visitors, though

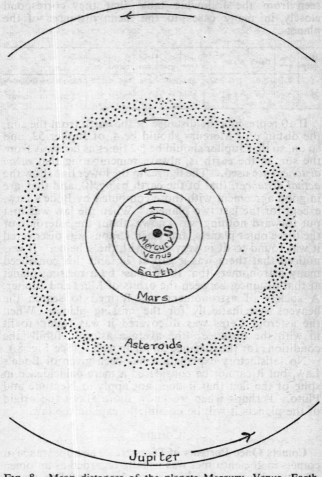

FIG. 8.—Mean distances of the planets Mercury, Venus, Earth, Mars and Jupiter from the sun are shown drawn to scale. The asteroids are shown scattered about between the orbits of Mars and Jupiter.

superstition has been largely replaced by the more rational fear of a collision with the earth—a fear not utterly baseless, though very often exaggerated. Unfortunately very bright comets are rather rare at present, and by far the greater number of those that come reasonably close to us can be seen only through the telescope. We use the word " unfortunately," because a bright comet is a wonderful spectacle and never fails to arouse awe and admiration in the minds of spectators. The last bright comet that we saw in the northern hemisphere was Halley's Comet in 1910, and it caused a considerable amount of interest in this and other countries.

Comets Members of the Solar System. Comets are no longer regarded as coming from outside the solar system; it is believed that they are members of the sun's family just as much as the planets and asteroids, and indeed the paths which they describe in their motion round the sun are very similar to those of some of the minor planets. In many cases they move in close to the sun and then go out to an enormous distance, taking a long time—often centuries—to complete one revolution. Halley's Comet, about which we have just said something, comes within 55 million miles of the sun at its closest approach and then recedes to the enormous distance of 3,300 million miles 38 years later, completing a revolution in 76 years. None of the planets moves in such an elongated orbit as this, and there are comets which have much more elongated orbits than Halley's.

The planets and asteroids all move round the sun in the same direction, but the comets are quite indifferent about the courses in which they move. Some are almost in the plane of the ecliptic, like the planets; some move at right angles to the ecliptic, and between these extremes others pursue their journey at various angles. Many, again, move in the reverse direction to that in which the planets move, and with different inclinations; taken on the whole, there are less with this retrograde motion than there are with direct motion. All this is very puzzling to the astronomer, who tries to find out how the planets and comets came into their present orbits, and up till now no one has been able to provide a satisfactory answer. (See Fig. 9.)

Composition of Comets. A comet consists of three distinct parts: the nucleus, the coma, and the tail. The nucleus consists of myriads of small particles varying in dimension from a good-sized house to specks of dust. It is a mistake to think that the nucleus is one solid body; it may contain a number of large solid bodies, and if any one of these struck the earth it would certainly cause some damage. The coma is a foggy-looking disc which surrounds the nucleus and enables us to distinguish the

FIG. 9.—The paths followed by a comet and a planet are shown in the diagram, the broken line denoting the comet's path. The planet moves nearly in a circle, while the comet follows an elongated ellipse, coming closest to the sun at P and going off to its greatest distance at A. Near the sun a comet develops a tail which points away from the sun, as shown. At other parts of its orbit the tail disappears. The comet is moving in a direction opposite to that of the planet. Less than half the comets have this retrograde motion.

comet from a star or a minor planet. The coma varies very much in size, but on the average we may compare it to the size of Jupiter. This is only a rough guide, as the coma may be ten times bigger than Jupiter, or even more. The coma of the great comet of 1811 was a million miles across, but this is rather exceptional. The tail of a comet develops only as the object approaches the sun, but there are comets which never develop tails, and there are others which develop two or more tails. The minute particles composing the tail—very much smaller than those in the nucleus—are driven off from the nucleus by a repelling action of the light of the sun, and this action increases as the comet comes near to the sun. Some of the light by which a comet shines is simply reflected sunlight, just as we see Saturn's rings by the light of the sun, which the particles reflect. This, however, does not explain *all* the light of the coma and tail. It has been shown that a great part of the light from these portions of a comet is due to glowing gas. Perhaps we are looking at something which is similar to the glow of the aurora in the atmosphere of the earth.

Comets' tails are sometimes very long—many millions of miles—and the earth occasionally passes through them without suffering any ill effects. This happened on May 19, 1910, when we went through the tail of Halley's Comet; the tail was estimated to be about 19 million miles long at the time. The tail is often the most frightsome-looking portion of a comet, but in fact it is the most harmless. The nucleus contains the possibilities of destruction if it encountered the earth, and though its mass is very small in comparison with that of the planets, the speed at which it would strike us would be responsible for serious consequences. It has been estimated that the least mass of the nucleus of Halley's Comet was 30 million tons, and it may have been much greater than this.

There has been much speculation about the origin of comets. Were some of them ejected by the larger planets, Jupiter and Saturn especially? Were they thrown out by the sun at a time of great activity on his surface? Are they the remains of debris which was torn out of the sun by a passing star and most of which condensed to form the planets? Are they visitors from outside the solar system? We can answer the last question now in the negative, but still we do not know where they came from, and it must be left to posterity to solve the problem.

METEORS OR SHOOTING STARS

Some Meteors are the Debris of Comets. There is a close connection between meteors and comets. This is illustrated in the case of Biela's Comet, which revolves round the sun in a period of $6\frac{3}{4}$ years. It was observed to split into two parts when it made its return near the earth in 1845, and at its next return in 1852 these two parts had separated much more. The comet should have returned again at the end of 1858, but it was not seen, and actually it was not very well placed for observation. Conditions for seeing it were more favourable in 1866, but again it was not observed. In 1872 it " appeared," in the form of a most wonderful display of meteors, on the night of November 27, and this shower has been repeated since at about the same date, sometimes as a considerable shower, sometimes as only a very feeble display. The debris of Biela's Comet had thus provided the material for producing a shower of meteors or shooting stars.

Size and Velocity of Meteors. Meteors are very small particles—most of them about the size of grains of sand—which are moving in orbits round the sun like the comets, and encounter the upper regions of the atmosphere in their motion. Their speed is about 26 miles a second, and as the earth is moving at a speed of $18\frac{1}{2}$ miles a second, it is possible for these bodies to enter our atmosphere with a speed of about 45 miles a second if they meet the earth direct. If, on the other hand, they are following the earth, the speed with which they strike the atmosphere will be the difference between 26 and $18\frac{1}{2}$—that is, $7\frac{1}{2}$ miles a second. In this case the earth's attraction hurries them on a little, and the actual speed is about 10 miles a second. Between the extremes of 45 and 10 there are intermediate speeds with which they enter the upper atmosphere, depending on the direction in which they are moving with reference to the earth. Their speed is responsible for their destruction. Friction with the atmosphere raises the molecules to a very high temperature, and the tiny bodies are rapidly burnt out, a flash of light informing us of their destruction. The heights at which they appear and disappear vary according to their speed and size, but generally they are seen first at a height of about 70 miles and disappear when they are 40 miles from the surface of the earth.

It must not be assumed that all the meteors which we see are the debris of comets, though there are several well-known meteor showers which are definitely associated with comets, some of which are still in existence and some (like Biela) quite defunct. An expedition organized some years ago by Harvard Observatory, U.S.A., to study meteors from observations made in Arizona, was responsible for certain views about these bodies being modified. It was found that most of them were moving much more rapidly than had been previously assumed, some as fast as 120 miles a second. Such a high velocity would imply that they were not members of the solar system but came from outside space. These results did not include those meteors which are accepted as associated with comets, but they included the great majority of meteors. In the last few years, however, much doubt has been expressed, with very good reasons too, on the validity of the results of this expedition. It is probable that most meteors are moving in orbits round the sun, and so are members of the solar system, though only a comparatively few of the well-known showers which occur about the same period each year are due to the debris of comets.

METEORITES

Occasionally some of the meteors are fairly large and are not completely burnt up by the heat which they generate in the atmosphere. In these circumstances the bodies or fragments of them rush through the atmosphere, sometimes with a loud noise and a strong light, and strike the earth. It may seem surprising that they do not penetrate the ground deeply, considering the high velocity with which they enter the atmosphere, but the resistance of the air reduces their speed very quickly. We do not witness the fall of many of these meteorites in a small country like England, but in America there are numerous records of the phenomenon, and there are many collections of the fragments of these bodies in the museums. Occasionally a meteorite of enormous size strikes the earth with great force and leaves a crater. In Arizona, near Cañon Diabolo, there is a crater in the desert, 570 feet deep and 4,200 feet in diameter, which is almost certainly due to the impact of a meteorite. Within a radius of 6 miles from the crater small pieces of meteoric iron, weighing several tons, have been collected, and these pieces were evidently once associated with the great mass which now lies buried in the earth. It is uncertain when the meteorite fell, but as cedar trees 700 years old are growing on the rim of the crater the event must have taken place before they started their growth. From the weathering of the rocks it is believed that the fall could not have been more than 5,000 years ago, and it may have been at any time between these extreme limits.

A very large meteorite fell in a forest in north central Siberia on June 30, 1908, and devastated an area of 3,000 or 4,000 square miles. The trees were blown down by the blast and lay with their tops pointed away from the centre of the area. This meteorite must have broken up before impact, or perhaps it consisted of a number of large bodies moving close together, because many craters have been discovered in the neighbourhood, the largest about 150 feet in diameter.

If any of these large meteorites fell in a populous district there would be terrible destruction of life, but fortunately such falls are very rare.

Questions

1. In what respects do the asteroids differ from the planets?
2. Why do you think that the asteroids were not formed in the same way as Saturn's rings?
3. State Bode's Law and show how it has been useful in finding planets.
4. What is a comet? If comets are dangerous, wherein does their danger lie?
5. In what respects are comets similar to the asteroids and in what special way do they differ?
6. Why is a comet's tail pointed away from the sun?
7. Explain why the coma and tail of a comet shine.
8. What comet broke up and produced a shower of meteors which can still be seen in November occasionally?
9. What are meteors? How high are they, on the average, when they appear and disappear?
10. Why do meteors glow and then disappear?
11. Distinguish between a meteor and a meteorite. Which is the more common?
12. State what you know about the Arizona Crater.
13. Why would craters formed on the moon by meteorites be more likely to survive than those formed on the earth?

CHAPTER VI

THE STARS

Distances of the Stars Expressed in Light-Years. Up to the present we have considered only those bodies which are members of the solar system, whose distances from us are measured in millions of miles or sometimes in the more convenient unit of the mean distance of the earth from the sun. This distance—93 million miles—is a useful " yardstick " or unit of measurement for many astronomical purposes, so long as we keep within the solar system. When, however, we pass beyond the bournes of our own little system to the so-called " fixed stars," the earth-to-sun unit becomes too small. It is then necessary to adopt the more convenient standard of a " light-year "—that is, we record the distance of the star in the number of years which its light takes to reach the earth. Light travels from the sun to the earth in about 8 minutes. The time required for light to travel from the nearest star to the earth is $4\frac{1}{3}$ years—figures which will give some idea of the immensities of space. One light-year represents nearly 6 million million miles, and a convenient way of expressing this is by writing it in the form 6×10^{12} miles. This is about 66,000 times as far away as the sun is, and it must be remembered that we are now considering the *nearest* star. Later on, when we come to deal with some of the far-off stars, the figures will appear truly staggering.

We have already seen that the planets move round the sun and that this motion can be detected by observing them for a short time against the background of the stars. The slowest-moving planet which is easily visible to the naked eye is Saturn, and if you watch this planet carefully through one month you will notice that it appears to move across the background of the stars for a distance which is about twice the diameter of the moon. On the other hand, you will not find that you can detect the movement of any star with reference to the others. They all appear to share equally in the movement from east to west, due to the rotation of the earth on its axis from west to east. There is, however, no visible change in their *relative* positions, so we speak of the " fixed stars." Nevertheless, all these stars in the heavens are in motion —some of them with speeds of several hundreds of miles a second—but they are plunged in the depths of space at such enormously remote distances that only the most delicate instruments of the astronomer can detect their separate movements.

How the Distances of the Stars are Measured. The distances of many of the stars have been measured, and it will be of interest to know how this is done. In a previous chapter a short description was given of the method of measuring the distance of the moon from the earth, and the same method has been employed to find how far the sun is away. In the sun's case, however, there are other ways of determining this distance more accurately, but they are too difficult for beginners to understand. It will be sufficient to recall that a base line and two angles are all that are required to find the other sides of a triangle, and the longer the base line the better the results. Suppose for instance, that we wanted to find how far away a landmark was and that, with a base line of 60 feet, we found the two angles between this base line and the line drawn from the object to each end of the line in turn to be 85° and 94°. The other angle of the triangle would be only one degree, and if we used an instrument which was not very accurate—say a prismatic compass—small errors could easily occur in measuring the angles, which would introduce very considerable errors in our calculations. This difficulty could be avoided by using a more sensitive instrument than a prismatic compass, but even with the most refined instruments there is a limit to the accuracy with which readings can be taken.

The other and more promising method is to increase the length of the base line. Suppose we took a base line of 4,000 miles on the surface of the earth and attempted to use it for finding the distance of the nearest star; it would be the equivalent of using a base line of 8 feet to find the distance of an object ten million miles distant. Needless to say, no accuracy could be secured under such conditions, and so any distances that we can measure on the earth are useless for finding the distance of a star. There is, however, another line which can be used, and this will now be described.

The earth takes a year to move round the sun, and its mean distance from the sun is 93 million miles. A base line twice this length—186 million miles—is thus provided for the astronomer if he waits 6 months for his observations. Even with such a long base line he must use the most delicate instruments to obtain anything approaching accurate results. The length of this base line compared with the distance of the nearest star is similar to a base line of 1 yard taken to find the distance of an object 26 miles off. This will give some conception of the accuracy which is necessary to obtain reliable results. The whole process of finding the distances of the stars is too involved for a simple explanation, but you can accept the figures as trustworthy, though, of course, allowance must be made for small margins of errors, especially in the case of stars which are far away from us.

Stars Vary in Brightness. You need only look at the heavens on a clear night to see that there are considerable differences in the brightness of the stars. If you examine them very carefully you will see, even with the naked eye, that there are differences in colour as well, though these are not so easily detected as the differences in brightness. The question arises, Why the differences in brightness? Are they due to the fact that some stars are much larger than others? Are they caused by differences in surface brightness, apart from size? You have often looked at two electric bulbs of the same size, and noticed a considerable difference in their luminosity. Then again, is it possible that the differences in luminosity arise from the various distances of the stars? The answer is that all three factors are at work, and one very important task which the astronomer has to undertake is to find out how much is due to each cause separately.

A detailed explanation of this fascinating branch of

astronomy would be quite impossible in a small work like the present one, and readers must accept the results which are given. Later, when they have mastered the elementary principles set forth in this volume, they may feel disposed to pursue the subject in more advanced works which will supply full explanations of the astronomer's equipment and of the results which he is able to obtain from its use. Here is one example of the sensitive instruments employed. It is necessary to find how much heat a star radiates to know something about its physical conditions, and very delicate instruments have been invented to do so. The heating effect of a star far beyond the range of naked-eye observation—in fact, visible only in a good-sized telescope—is about the same as the heat that we should derive from a candle 2,000 miles distant. Yet this heat can be measured by a "thermocouple" used in conjunction with a large telescope!

Magnitudes of Stars. It is convenient to have some standard of brightness when we are referring to different stars, and the method in universal use is as follows:—

A star of magnitude 1 differs in brightness from a star of magnitude 2 in such a way that the brightness of the first is 2·512 that of the second. *It is important to remember that the smaller the number denoting a star's magnitude, the brighter is the star.* Similarly, a star of magnitude 2 is brighter than one of magnitude 3 in the same ratio, so if we could imagine 2·512 stars of magnitude 3, the combined brightness would be the same as that of a star of magnitude 2, and so on. When we compare stars of different magnitudes which are not consecutive, say a star of magnitude 1 with a star of magnitude 3, we take the difference between 3 and 1 (which is 2) and raise 2·512 to the second power—that is, square 2·512. The result, 6·31, tells us that a star of first magnitude is 6·31 times as bright as a star of magnitude 3. The same would apply in other cases—that is, a star of magnitude 4 would be 6·31 times as bright as a star of magnitude 6, and so on. Now suppose we want to know how many times brighter a star of magnitude 1 is than a star of magnitude 6. Subtract 1 from 6 and the result is 5; multiply 2·512 by itself four times (usually expressed in the form 2·512⁵), and it will be found that the first-magnitude star is just 100 times as bright as the star of sixth magnitude. This shows that there is a certain amount of simplicity in the scheme, because it is easy to remember that a first-magnitude star, of which there are twenty (and fourteen of these are visible in our island), is exactly 100 times as bright as the sixth-magnitude star, which is just visible to the naked eye.

Sirius, Rigel, Capella, Vega, Betelgeuse, Aldebaran, Spica, Pollux, and Regulus are some of the best known first-magnitude stars, and you should learn their positions with the aid of a star map. Brighter stars should be known first, and then later on the fainter stars can be learnt by noticing their positions with reference to the conspicuous stars.

Sizes and Densities of Stars. Dealing with the first factor referred to as making some stars look brighter than others—the fact that some stars are very much larger than others—many results which have been obtained in comparatively recent times are interesting. It has been found that there is an enormous difference in the sizes of the stars, but, strange to say, our sun, which is just a star like the others, is a fair average in size as well as in mass and temperature. We have already shown that the sun's diameter is 864,000 miles, and this will be a useful unit for comparing the diameters of some of the other stars. The mass and density of the sun will also be taken as units, and those of the other stars will be expressed in terms of the sun. Thus, if the density of a star is given as 0·0021, it means that this is its density if the density of the sun is 1. Suppose we want to find the star's density in terms of the density of water. We know that the sun's density is 1·41 times that of water, and hence the density of the star would be 0·00296 that of water. The same method applies to the mass. If this is given as 4, it means that the mass is four times that of the sun, and as this is 2×10^{27} tons, the mass of the star would be 8×10^{27} tons.

One very remarkable thing is that, although the sizes and densities of stars vary very much, yet the differences in their masses are not very great. Not many stars have been discovered with a mass more than ten times or less than one-fifth that of the sun, but there are exceptions to this.

Name of Star	Diameter	Density	Mass
Antares	480	0·00000021	30
Aldebaran	60	0·000014	4
Arcturus	30	0·00021	8
Capella A	12	0·0014	4·2
Sirius B	0·034	19,000	0·96
o_2 Eridani B	0·019	45,000	0·44

Some of these figures present interesting features. Let us examine those of Antares at the beginning of the list. Perhaps the first thought which will occur to readers is that, as it is so large it must be intensely hot, but this is not correct; in fact its surface temperature is about half that of the sun. Its small density is the most remarkable thing about it. It is difficult to form any conception of this low density—about one three-millionth of the density of water. Water weighs 773 times as much as air at sea-level, so that the density of the star must be about one five-thousandth part of the density of air. In contrast with this very low density, its diameter is enormous—about 415 million miles. We have seen that the greatest distance of Mars from the sun is 141 million miles, hence it would be possible to place our sun and Mars in his present orbit inside Antares, the sun being at the centre, and there would be more than 60 million miles to spare before reaching the outside of the star. It may be remarked that Antares is about 380 light-years from us, and it ranks amongst the first twenty brightest stars in the heavens.

The second star, Aldebaran, is a little brighter than Antares, but is only one-sixth of the distance of Antares. We should expect that it ought to be much brighter because it is so much nearer to the earth, but this is one illustration of the fact that the brightness of a star does not always depend on its distance from us. While the rule that brightness falls off with distance holds in general, yet there are many exceptions, because, as we have seen, size and surface brightness are also important factors in determining the brightness of a star.

White Dwarf Stars. At the end of the list are two stars which have densities beyond anything that we ever experience on this earth. Sirius B has a density 19,000 times that of the sun, or about 27,000 times that of water —a remarkable contrast with Antares, which is like a gas bubble. Although the diameter of Sirius B is only one-thirtieth that of the sun, and hence the volume is one-twenty-seven-thousandth that of the sun, the mass is nearly the same. It is easy to show that with such a density a cubic foot of the material would weigh 750 tons (a cubic foot of water weighs 63 pounds), and it would be impossible for a man to lift even a cubic inch of the material of this star. The figures refer to the *mean* density; actually the density increases as the centre of the star is approached. Sirius B is known as a " white dwarf," the first word denoting its colour and the second its size. Its high density arises from the crushed state of matter which composes it, and it is believed by some that this will be the fate of the stars in general, including the sun. A discussion of this would lead us too far in an elementary treatise, and we must now look at some other peculiarities of the stars.

It may have occurred to the reader that the enormous stars and the small stars which have just been referred to merely represent different stages in the evolution of stars. This conjecture is correct.

Stellar Evolution. In the table on the previous page the star Capella A is included in the list. This star has a diameter 12 times that of the sun, and its surface temperature is about 1,000° C. less than that of the sun. It is known as a red giant, and a characteristic of this class is that it possesses a large diameter and low surface temperature, as well as a small density. The temperature in the interior of these red giants, as with all stars, is greater than on the surface, but it does not attain the value of the temperature in the interior of our sun or other similar stars. Thus it has been estimated that the temperature at the centre of Capella A is only 5 million degrees Centigrade, which is about one-fourth the temperature in the central regions of the sun. It is believed that each star begins its life as a giant globe of gas which is very rarefied and cold, but its own gravitation causes it to contract and to increase in temperature. When the contraction has proceeded for some time the central temperature may rise to a million degrees, and owing to the heat energy released in the interior the contraction is stopped. After a certain period, however, the output of energy has diminished to such an extent that contraction begins again, and then, when a higher temperature has been attained, it ceases once more. The process is continued until the temperature of the star is high and its luminosity is considerable. During this process the star is continually transforming its mass into radiation of heat and light,* so that the older a star is, the greater the amount of matter it has transformed in this way. It has been estimated that the sun is losing his mass at a rate of 4,000,000 tons every second, and though this seems an enormous amount, it is extremely minute when compared with the mass of the sun. Stars with different masses run through their evolutionary lives at different speeds ; the heavier a star is, the sooner it will reach maturity, old age, and final death, so far as emission of heat and light is concerned. After attaining a high luminosity, due to increasing temperature through contraction, at which stage the star is a giant, the final contraction begins and eventually leads to the death of the star.

The white dwarfs, of which Sirius B and Eridani B are typical, present certain difficulties. Not many of these remarkable stars are known, but this does not necessarily imply that they are actually scarce. Owing to the fact that they emit a very feeble light, it is difficult to detect

* When we burn a piece of coal the weight of its ashes and the smoke emitted is very nearly the same as the original weight of the coal, but not quite. To make up the original weight it is necessary to add in the weight of the light and heat emitted during combustion. This analogy explains the meaning of the transformation of some of the sun's mass into radiation of heat and light. The weight of the oxygen which enters into chemical combination is ignored, to avoid complicating the analogy.

them unless they are fairly near us. Within recent years Luyten and Kuiper have made a special study of these bodies and their list now includes about forty white dwarfs. It may be presumed that there are others at greater distances which will escape discovery for many years and perhaps a large number will never be discovered.

Sir James Jeans has informed us that the sun is perilously near the stage where it might collapse into a white dwarf and the diminution of its light and heat would imply the end of all life on the earth. Nevertheless, although a reduction of only three per cent in the sun's luminosity would place our luminary on the edge of the precipice where it would collapse, we are assured that this cannot happen for at least 150,000 million years. In addition, the sun is not heading straight for the precipice, but is skirting it, so on the whole we have a respite for another million million years. Professor G. Gamow, George Washington University, presents us with a similar picture of our future, but before the sun collapses into a white dwarf he thinks there will be an enormous increase in the amount of solar radiation, which will make life very difficult if not impossible on our planet. However, the slow rise in temperature will probably be accompanied by evolutionary changes in the biological world, and some terrestrial life will be able to adapt itself to the increasing heat. He does not believe that the higher species will be able to accommodate themselves to the increased temperature, so that the last radiation efforts of the sun, after its output of heat is succeeded by the collapse into the white dwarf stage, will be observed only by the simplest of micro-organisms.

The prospect is not very enticing, but the present writer's opinion is that millions of years before there is any noticeable alteration in the amount of the solar radiation, *Homo sapiens* will have become extinct like hundreds of other species which once flourished on this planet. From the point of view of a member of the human family perhaps he should express some regret at the fate of the species to which he belongs, but from the cosmic point of view there is little or nothing to deplore in this final issue.

Questions

1. If you were uncertain whether an object was a planet or a star, how would you decide by naked-eye observations?

2. Why is it impossible to determine the distances of the stars in the same way as the distance of the moon is found?

3. Why do stars differ in brightness?

4. In what respects do some stars differ very much from the sun, and in what special way is there an average relationship between the sun and other stars?

5. Describe briefly the course of stellar evolution.

6. What is the difference between a red giant and a white dwarf?

7. What, according to the most recent research, will be the ultimate fate of the sun?

8. What possible alternatives await (*a*) human life, (*b*) all forms of life on the earth?

9. Why is the title "fixed star" applied to the stars when none of them is at rest? In what respect can you consider them to be fixed?

10. Close to Castor, magnitude 2, is the star ρ Geminorum, magnitude 5. How many times is Castor brighter than ρ Geminorum?

CHAPTER VII

THE STARS (*continued*)

Double Stars. To the naked eye every star is a single point of light, but even a small telescope will show that many of them have one or more companions close to them. It must not be inferred that these other stars which appear in a telescope to be close to the brighter one are necessarily connected with it. They may just happen to lie nearly in the same line of sight, but one may be scores of light-years farther off than the other. In such circumstances the members or components are called " optical doubles." Interesting as they may be, they are of little or no scientific importance. There is, however, another class of double stars which presents the astronomer with some of the most interesting and valuable problems in celestial mechanics. Not only do these stars seem near to one another; they are actually fairly close and they are revolving round their common centre of gravity just as the earth and moon do.

These double stars are known as " binaries," and their period of revolution varies from a few days, when they are very close together, to hundreds of years when they are widely separated. Not only are there binaries which revolve round the common centre of gravity; triplets and quadruplets and even greater numbers have been found which do the same thing, and these present a problem which the mathematician has never been able to solve. It is not a very difficult thing to explain the revolution of two bodies around their common centre of gravity and to predict exactly where each one will be at a certain time. It is impossible, however, to do the same in the case of three bodies, unless one of them has a very small mass, and of course it is equally impossible to tackle the problem when there are four or more. Here is a case where phenomena in the universe baffle the mathematician and probably will continue to baffle him.

In the previous chapter reference was made to Sirius

B, the companion to Sirius. Many years before this companion was discovered it was known that it was there, because Sirius was observed to move in a wavy course —a clear hint that it was revolving round some centre. Twenty-eight years later the companion which had been suspected was found by Alvan G. Clark, a telescope-maker, who was testing a new refractor.

Many double stars present a beautiful appearance even with a small telescope, and their beauty is sometimes enhanced by the diversity in the colour of the components. If you have the opportunity to look at ε Lyrae you will see that it is a quadruple system. Sometimes there are six components, like θ Orionis in the sword of the well-known constellation of Orion. If a planetary system existed in connection with a binary or a system of three or more stars, the movements of the planets would be very complicated and it would require superminds to predict where a planet would be at any time.

When the time of a revolution is known from observing the motions of the components over a sufficiently long period, and we also know the distance of the system from the sun, it is possible to measure the combined masses of the stars. In this way many stars have been weighed. Their masses have been found to differ widely, though there are decided limits. One system has been found in which the masses of the components are at least 75 and 63 times the sun's mass, but this is very exceptional, and it is almost certain that few such systems exist. On the other hand, there is a lower limit to the mass of a star. The faintest member of the triple system o_2 Eridani has a mass only one-fifth that of the sun, and this is the smallest mass in any star yet discovered.

Research on binaries is a specialized branch in astronomy, and there are many very important problems arising out of it, some of which are concerned with the ages and origin of the stars. It would be impossible to deal with these at present, and we shall now proceed to consider another kind of star which also presents fascinating problems, some of which still await solution.

Variable Stars. Eclipsing Binaries. Those who have

not made a study of elementary astronomy are under the impression that each star always emits the same amount of light, and that variations in the brightness of a star are due to atmospheric conditions. This is not correct, however, and there are a great many stars whose magnitude change, some in a regular manner, others in ways that seem erratic. There is one star in particular which can be observed to change rapidly in brightness in the course of about 3 hours. The sudden fluctuations in the light of this star were known to the Arabians, who called it Algol, meaning the " Demon Star." It is still known by the name Algol, but a more scientific designation is β Persei. Its phases are briefly as follows: In about $3\frac{1}{2}$ hours it descends by two magnitudes in brightness and remains for 20 minutes like this ; then it begins to brighten until it regains its original condition after another period of $3\frac{1}{2}$ hours. For $2\frac{1}{2}$ days it remains bright, and then the changes begin again and run through the same phases in a most regular manner.

The previous section dealt with binary systems, and the reader may now conjecture the reasons for the variations in the light of β Persei. Suppose that one of the components is not so bright as the other or that it is almost a dark body, and also that the plane of the orbit in which the stars revolve is edgewise or nearly so to us (see Fig. 10). In these circumstances some of the light of the brighter body will be cut off when the other component comes between its companion and the earth. This is what happens in the case of β Persei and other " eclipsing binaries "—the name given to this class of variables. If the plane of the stars' orbit is not edgewise to us there will be less shutting off of the light than in the case where it is absolutely edgewise, hence eclipsing binaries will not always show the same amount of change in brightness. There are other reasons why the changes should not be the same for different systems—among which the relative size and brightness of each component are important. Five more of these eclipsing binaries can be observed with the naked eye, and probably the most convenient of these for studying is β Lyrae. The interval between its times of minimum brightness is nearly 12 days 22 hours. Many eclipsing binaries exist whose changes in brightness can be detected only with the delicate instruments at the disposal of the astronomer.

Long-Period and Irregular Variables. There are several other classes of variable stars, and we shall deal with the long-period variables and the irregular variables —names which suggest the characteristics of these classes.

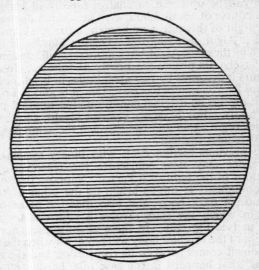

Fig. 10.—Showing an eclipsing variable. Hold the paper before the eye which can be taken as the position of the earth. The brighter star has almost completely passed behind the darker one and hence there is a marked falling off in its light.

The long-period variables are red giants, of which something was said in the previous chapter, and their periods range from about 2 months to 2 years. The most common period is in the neighbourhood of 300 days. One well-known example of this class is o Ceti, generally known as Mira, "The Wonderful," which varies from magnitudes 3·5 to 9 in an average period of 330 days, but this time may alter by as much as a month. More than 1,700 long-period variables have been catalogued, but only about twenty of these are visible to the naked eye.

Irregular variables are also red giants, but they differ from the long-period variables in certain respects. First of all, it is absolutely impossible to predict what they will do. In the next place, they do not alter to the same extent as the long-period variables, their variations seldom exceeding half a magnitude, and very often they are much less than this. A well-known star of this class is Betelgeuse in Orion.

The Cause of Long-Period and Irregular Variables. The astronomer is not content merely to record what he observes ; he seeks for an explanation of the phenomena, and so he has tried, not with complete success, to explain the cause of the long-period and irregular variables. A theory to explain these and also the Cepheid Variables (see next paragraph) is as follows.

A large gaseous star contracts under its own gravitation, and in doing so becomes hotter. A familiar example of this kind of heating is found in a pump for a car or bicycle. Compressing the air inside the pump causes the air and the pump to grow warmer, as anyone can verify for himself. Again, when a gas is heated it tends to expand. If anyone doubts this he can hold a child's toy balloon or a blown-out football bladder close to a fire ; the pressure from the heated gas which is trying to expand will quickly burst the balloon or bladder. In the same way the heated gas in the contracted star will produce an expansion, and this in turn will cause a certain amount of cooling. When cooling has gone to a certain stage, contraction will commence again, and the process will be repeated. This action has been likened to the beating of the heart. During the expansion and contracting of these stars it is thought that the diameters may change by as much as 10 or even 15 per cent, while the range of temperature may be 1,000° C. In the expansion stage the star becomes brighter, and as it contracts, although it is hotter, it becomes smaller, and hence gives less light. While this " pulsation theory " is open to certain objections, it provides a tentative explanation of the varying brightness of these stars.

Cepheid Variables. This class of variables takes its
name from δ Cephei, one of the earliest recognized
examples. Their periods vary between a few hours and
about 50 days, and, with a few exceptions, their light
variations are continuous and remarkably regular. The
distances of these stars can be determined when their
apparent magnitudes and lengths of periods of light
variations are known, but an explanation of the method
is beyond the scope of this work.

Sir James Jeans thinks that the variation in their
light arises from rotational instability. His theory is
that the stars are rotating so rapidly that they are on the
point of breaking up, just as a weak fly-wheel might
burst if its rotation were very rapid. Some stars are
rotating rapidly, and so their equatorial regions bulge
to an enormous extent. In an earlier chapter we saw
that Jupiter's equatorial diameter exceeded his polar
diameter by more than 6 per cent, but in the case of the
variables under consideration the equatorial diameter
may be twice as long as the polar diameter. Alternate
expansion and contraction of this long axis in combination
with the rotation explains some of the phenomena of
these variables, but other peculiarities are not explained
satisfactorily in this way, and much still remains to be
done to account for the light changes in this type of stars.

Novae. In addition to the variables which have been
considered, there is another type of star which might be
described as a variable, but which differs so much from
the other stars of this class that it has a special title of its
own—Nova or Temporary Star. These stars rise very
suddenly from obscurity to a great brightness and then
sink back again into faint stars. Five brilliant novae
and also a number of dim novae have been discovered
in the present century. The last bright one was discovered
early in 1934 by Mr. J. P. M. Prentice, Director of the
Meteor Section of the British Astronomical Association,
who was looking out for meteors at the time.* This

* Just before the MS. was sent to the printers, information was
received that a nova was discovered by Mr. T. Ellis, Llandudno,
North Wales, on the morning of November 13, 1942. Its magnitude
at the time was about 2 and it was in the Constellation Puppis.

nova, which is known as Nova Herculis, from the name of the constellation in which it was discovered, flared up to brighter than 2nd magnitude. Photographs which had been previously taken of this region of the heavens showed that there was a star where the nova appeared and that its magnitude was 15. A calculation which can be made from the data supplied in the previous chapter shows that this star must have increased its luminosity 160,000-fold in the course of a few days.* The sudden change indicates that the surface of the star must have got enormously hotter in a very short period, and suggests that there must have been something like an explosion. In some cases the speed of the gases surrounding a nova has been shown to exceed 1,000 miles a second, and observations show that successive surface layers are blown off at intervals of a few days.

An interesting fact in connection with Nova Persei, discovered in 1901, is worth mentioning. A few months after its outburst a faint glow was photographed around the star, and this was seen to increase at a rate which was later found to be the speed of light. This did not imply that the material blown out of the star was receding with this speed. It showed that the star was within a region containing dark cloudy matter which was illuminated by the sudden light produced in the nova. This light spread outwards from the star with a speed of 186,000 miles a second and illuminated the previously dark nebula, which then seemed to an observer on our planet to be expanding with the velocity of light.

Cause of Novae. The precise cause of a nova is not known, but we are fairly certain that the possible causes can be narrowed down to a very few. A collision with another body has been suggested, but the chance of this is too small to account for all novae, including those which we can see directly and more which the camera detects. A more likely theory is that the star collides with a cloud of nebulous matter, producing an enormous amount of heat, just as would happen if a collision between two bodies occurred. We have already seen that meteors are heated to very high temperatures when

* $2 \cdot 512^{13} = 160,000$ approximately.

they collide with our atmosphere, and the same thing would probably occur if a star collided with a nebula. Another theory—and the most probable of all—is that the nova is produced by something within the star itself. There is a sudden release of energy which is responsible for a high surface temperature and the rapid increase in brightness. It is very improbable that anything more than the surface is affected because the novae generally collapse into faint stars very soon after the sudden flare up, and this would not occur if the greater part of the star had been affected.

Perhaps every star passes through the nova stage at some time in its history. If this is so the sun may become a nova, in which case the inhabitants of a planetary system on some far-off star, if such planetary systems and inhabitants exist, would announce the discovery of a nova. Needless to say, inhabitants of this or any other planet in the solar system would not be aware that a nova had appeared, because the whole planetary system would be reduced to the gaseous state in a very short time through the intense heat produced at the time of the outburst.

Questions

1. What is the difference between an optical double and a binary?

2. Is it possible by naked-eye observations alone to distinguish between an optical double and a binary?

3. What important deductions can be made from a binary system when the time of revolution and the distance from the earth are known?

4. What star can be observed with the naked eye to vary considerably in brightness?

5. Explain how an eclipsing binary causes variations in the amount of light which reaches us. Use a diagram to illustrate your explanation.

6. How do long-period and irregular variables differ from eclipsing binaries? To what class of star do the former belong?

7. Give a short explanation of the two main theories which have been advanced to account for long-period and irregular variables, and Cepheid Variables.

8. A nova is a variable star, but differs entirely from the usual type of variable. Explain in what respects it differs.

9. What is the cause of a nova? Why do you think that a collision between two bodies is not the cause, generally speaking?

CHAPTER VIII

NEBULAE

The Milky Way System or Galaxy. The Milky Way is one of the most impressive sights in the heavens, and probably ancient astronomers were puzzled to explain it. It is remarkable, however, that Democritus, who lived in the fifth century B.C., suggested that it was made up of an enormous number of faint stars which could not be seen individually. It was impossible to confirm or deny this speculation until the invention of the telescope, and then it was shown that Democritus was correct in his guess. It is not necessary to use a telescope to notice that the stars in or near the belt are more numerous than those in other parts of the sky, and in fact all the stars that we can see with the naked eye belong to the Milky Way system. The term " galaxy " is usually applied to this system ; it is derived from a Greek word which means milk, and is a very appropriate title for this great girdle of the heavens.

The story of the research which astronomers have conducted on the galaxy since the days of Sir William Herschel is an extremely interesting one, but we have no space to deal with it. All that can be done now is to give the results of many years of patient work on the part of a number of astronomers who have devoted their time to a special study of this portion of the heavens. Owing to the vast distances which separate the extreme portions of the galaxy from one another, it is necessary to express distances in terms of light-years. Later on the dimensions will be given on the scale of a small model which will assist in visualizing the immensity of this system of stars in which our own sun is situated.

The galaxy is shaped somewhat like a bun, but is more flattened than most buns are. Imagine a bun which is about six times as long as it is deep, and you will have a rough conception of the shape of the galaxy. Inside this bun-shaped system there are about 3×10^{10} stars—

that is, 30,000 million—which are of various sizes and masses, but which, taken on the average, do not differ very much from one particular star that is important for us—our sun. Light would require about 120,000 years to cross the galaxy from end to end and about 20,000 years to cross its shortest diameter—referred to as the depth of the bun. We might expect that an important star like the sun (important at least from our point of view) would be near the centre of this vast system, but no special preference is shown to it, and in fact it is 30,000 light-years from the centre (see Fig. 11). Outside the galaxy there is a vast emptiness, and then we come to more similar galaxies, one of the nearest of which, the Great Nebula in Andromeda,* is nearly a million light-years away from us. This is very close in comparison with some others, which are as far as 140 million light-years away.

On a planet which is part of the system of an ordinary star situated 30,000 light-years from the centre of one galaxy out of 2 million, something which we call life has developed. It has reached its highest intelligent form, so far as we know, in a species Homo sapiens. This species has looked out on the universe and has meditated on the mysteries of its origin and destiny. It has weighed the stars and has analysed the elements composing them. It has plumbed the depths of space and has discovered that its own little planet, and even the whole planetary system of which its own planet is a part, are but a speck in creation. It has mused on the possibility of life elsewhere in the vast cosmos and has even tried, without success, to communicate with beings, if such there be, on other worlds. At one stage of its history it was so egotistic that it believed all creation was subservient to its needs and that the myriads of heavenly bodies were specially designed to minister to its wants.

Four hundred years ago, when Copernicus showed

* The nebula M 33 in Triangulum is slightly nearer, its distance being 850,000 light-years, whereas that of the Great Nebula in Andromeda, M 31, is about 900,000 light-years away. The Magellanic Clouds are very much nearer to us than either of these nebulae, but there is some doubt whether they are outside our galaxy or are part of our galactic system.

that the earth and planets were moving around the sun, the earth was dethroned from its central position in the universe, though man has been very slow to learn the lesson. Until recent times he could at least boast that the galaxy to which the sun belonged was unique in dimensions, but now it seems that there is nothing special about it, and even a near neighbour in galaxies—the Great Nebula of Andromeda, once believed to be very much smaller than our own—is comparable in size with the Milky Way. The largest telescope in the world, the 100-inch reflector at Mount Wilson Observatory, shows that there are many apparently faint nebulae, all galaxies, just at the limit of vision, which is 140 million light-years. There is no reason to suppose that the nebulae end there, and when the 200-inch telescope is completed it will penetrate the depths of space to twice this distance. The most remote nebulae that can be photographed with the 100-inch telescope are about 500 million light-years distant. The 200-inch telescope will probably penetrate to twice this distance. If the nebulae are distributed as uniformly in those far-off regions as they are nearer to us, then we must expect that 16 million will be discovered.

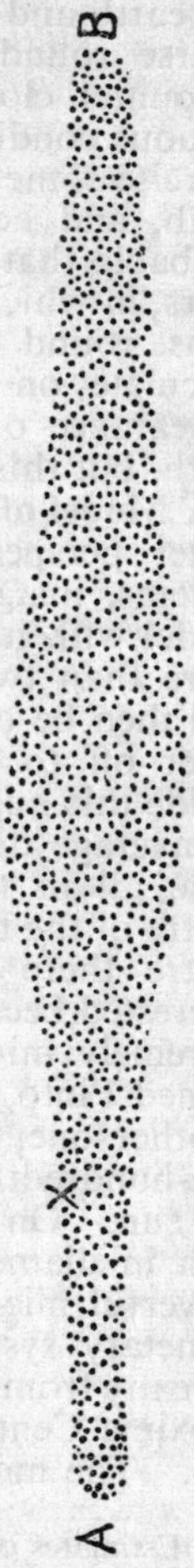

Rotation of the Galaxy. Every galaxy, including our own, is rotating, but not like a solid disc or a wheel. Different objects and classes of objects in the galaxy have different rates of rotation. The sun is moving around the centre of the system (which is in the direction of the dense star-clouds in the constellation of Sagittarius) with a speed of 170 miles a second, and completes a revolution in about 225 million years; of course, he takes with him

Fig. 11.—A model of the galaxy. The sun is marked x.

the earth and all the other planets and satellites. In his course round the centre of the galaxy the sun may encounter clouds of cosmic dust, or matter in a very tenuous condition, and it is possible that this dust could exercise some influence on the climatic conditions of the earth, and hence on the various forms of life. It is probable that the dawn of life dates back 1,000 million years, in which time the sun made four complete revolutions round the centre of the galaxy. One could speculate on the causes of the appearance and disappearance of many species which once inhabited the earth, but this must be left to the biologist.

A Model of the Galaxy. In order that we may gain a better perspective of the vast dimensions of galactic systems, imagine that our own galaxy is reduced to a model with its greatest diameter * 1,000 miles, which is more than twice the length of England. Our " bun " will then be only 160 miles thick at the centre, and will taper off to about one-fourth of this near the ends. Inside this huge bun there are 30,000 million specks with an average diameter of one ten-thousandth part of an inch, all in rapid motion at various speeds round the centre of the bun. Somewhere about 300 miles from the centre there is a particular speck in which we are interested because it is our sun. Around it a number of extremely minute specks are revolving, the outer one, named Pluto, at a distance of two-fifths of an inch, and another one, which we call the earth, at a distance of one-hundredth of an inch, from the large speck which is the sun. On this scale the earth is one-millionth of an inch in diameter, and so would be invisible even in a powerful microscope. About 80 yards away from this planetary system (it makes no difference whether we measure from the sun, the earth or Pluto) there is a star— Proxima Centauri—which is the next large speck to the sun. The most distant speck which is inside the bun is

* Estimates of the size of the galaxy and of the number of stars which it contains vary very much. The model is based on a diameter of 100,000 light-years, which is probably less than the actual diameter. The number of stars has been estimated to be anything between 30,000 and 100,000 million.

about 800 miles away, and then, when we go outside it, there is a vast void for 9,000 miles before we come to the next bun, which is M 33 in Triangulum. This is our nearest galactic neighbour, and it is comparatively close, since some have been detected with powerful telescopes more than $1\frac{1}{2}$ million miles away on this scale. Bear in mind that the earth is one-millionth of an inch in diameter according to the model of the galaxy. We see the most distant galaxy by the light which it emitted 140 million years ago. Many of the stars of this far-off system may have exploded and become novae for a few months, only to collapse again into faint stars. Perhaps a few may have collided and have ceased to exist as stars any longer, being turned into huge volumes of gas, while planetary systems with their inhabitants have perished like a moth in the furnace. Our telescopes are not sufficiently powerful to notice such trivial matters as these, and if they could reveal these cosmic upheavals it would be 140 million years after they occurred—probably three or four hundred times the period during which human life has existed on the earth. Such are the distances and the time scale which are used in dealing with cosmic problems.

The Spiral Nebulae or Extra Galactic Systems. The galactic systems far beyond the outlying stars of our own galaxy are called spiral nebulae, although many of them show no signs of the spiral form. If the plane of a flattened system is perpendicular to our line of sight it is easier to detect the spiral form than if it is edgewise to us. In the case of the Andromeda nebula the plane is inclined at an angle of 15° to the line of sight, and this is sufficient to reveal its structure. Towards the centre of the spiral the substance of the nebula is denser than the arms of the spiral. Owing to this feature the nuclear region of the Andromeda nebula can be seen with the naked eye, and no other spiral nebula can be seen without telescopic aid. Some nebulae are turned with their planes practically at right angles to the line of sight and exhibit the spiral form, including two great spiral arms very clearly, as in the case of M 51 in Canes Venatici. If our Milky Way could be viewed from a suitable position outside our system it would probably present the appearance of a spiral nebula, like other galaxies. We have seen that our galaxy is in rotation, and the same applies to the spiral nebulae. Measurements of the rate of rotation of M 31 suggest that 19 million years are required for a complete rotation, and this implies that the outer portions must be moving at hundreds of miles a second. It is fairly certain that a galactic system must have been produced out of a rotating body, or, rather, a gaseous mass.

The Spiral Nebulae are Running Away from One Another. A remarkable fact has been brought to light in recent times regarding the motions of the spiral nebulae. They are all running away from us and from one another, but not at the same speed, since the more distant ones are moving faster than those nearer to us. It has been found that a nebula at a distance of $3\frac{1}{2}$ million light-years has a speed of about 330 miles a second, and if the distance is doubled, the speed is doubled too; if the distance is halved, so is the speed. This extraordinary flight of the spiral nebulae involves some very difficult problems on which astronomers are not yet agreed, and the reader must be content to accept the fact that the nebulae are running away from us and from one another, without inquiring into the reasons for this strange behaviour.

Nebulae Which are Part of our Galactic System. It is rather unfortunate that the name "nebulae" should have been applied to the enormous stellar systems far beyond the Milky Way, as well as to a class of objects in our own galaxy which are utterly unlike stellar systems. In appearance they are similar; both are like a wisp of mist or cloud (the word "nebula" is the Latin for a cloud, mist or fog), but there the likeness ceases. Two types of nebulosity are found in the galactic system. There are the *diffuse nebulae*, of which the Great Nebula in Orion is an example; the luminosity of these is not due directly to a multitude of stars, as in the case of the spiral nebulae, but indirectly to the influence of a few stars lighting up dark matter in space. When you look at the Milky Way you will notice that it is not all equally

studded with stars; in some parts there are clouds
between us and the stars beyond, and this obscuring
matter cuts off the light which would otherwise reach us.
The luminosity of diffuse nebulae is due mainly to the light
from associated stars; the light of the stars is reflected
by the fine dust, or perhaps very rarefied gas, which
otherwise would be merely dark matter. In addition to
the reflection of the light of stars, it is certain that the
very hot stars are able to excite some of the gases in a
nebula to glow, and so we see diffuse nebulae not so much
by reflected light, as by light which they emit under the
stimulus of very hot stars. The Great Nebula in Orion,
in which the star θ Orionis is situated, is about 600 light-
years from us—quite a short distance in comparison with
the nearest spiral nebula.

Planetary nebulae present a different appearance from
the diffuse nebulae. As their name suggests, they are like
a planet, inasmuch as they are round, or nearly so, and
they are smaller than the diffuse nebulae. One of the
most beautiful of all the planetary nebulae (of which
seventy-eight can be seen in the northern sky) is the Ring
Nebula in Lyra. As we might infer from its name, it resembles a ring in its luminous portion. The central
regions are dark owing to obscuring matter. The
diameters of about twenty planetary nebulae have been
found, and they are several thousand times that of the
solar system. In most cases they have a white-hot star at
their centre, and this star is responsible for imparting light
to the whole nebula. It is certain, however, that it does
not shine by reflected light as the moon does, because the
nebula is brighter than it could be if reflected light alone
were responsible. In some way the light of the very hot
star is able to start the gases shining, as happens with the
diffuse nebulae, but the actual process still remains in
doubt.

How the Constitution of Celestial Bodies is Known. Up
to the present nothing has been said about the method
which is used to distinguish between the different kinds
of nebulae, especially between those which are outside
our system and those which are part of our galaxy. The
instrument which the astronomer employs for finding the

chemical composition of the stars and nebulae is known as a spectroscope, and it is simply an apparatus for studying a spectrum. When ordinary sunlight is passed through a prism it is broken up into its constituent parts, consisting of all the colours of the rainbow—violet, indigo, blue, green, yellow, orange, red—and the strip of coloured light which the prism throws on a sheet is known as the spectrum of sunlight. Beyond both the red and the violet ends of the visible spectrum there are rays which we cannot see. These are the infra-red (below the red) and the ultra-violet (beyond the violet) rays, the latter being particularly effective in photographic work.

Sunlight is not, of course, the only thing that produces a spectrum, or at least a portion of one. When any substance is made white-hot and its light is examined by a spectroscope (which is simply a combined prism and telescope, with a lens to make parallel rays from the white-hot substance fall on the prism), the spectroscopist, by examining the spectrum, possesses a lot of useful data. Every element has its own peculiar spectrum of bright and dark lines, and the spectroscopist knows exactly what each line means, so he has a mine of information at his disposal when he studies the spectra of stars and nebulae. He knows whether the light which comes from those bodies, separated from him by millions of light-years, is produced by a star or a glowing gas, and he can tell us a lot about the elements which go to make up the star. These elements, it may be remarked, are similar to the elements which he finds in the sun, and these again are similar to those found on the earth.

Another very important thing the spectroscope can do is to tell us the speed with which a body—star or nebula—is approaching the earth or receding from it. This is done by noticing how far certain lines in the spectrum are displaced from their usual position in which they are found when the object under examination is at rest. We have already referred to the gases methane and ammonia, which largely compose the atmospheres of some of the giant planets. It was by using the spectroscope that these gases were detected. It would be a very long story to tell all that the spectroscope has done, and those who are sufficiently interested in this branch can read more advanced or specialized works, which will explain how so many secrets have been wrested from the universe by this wonderful equipment of the astronomer.

Celestial Photography. The camera is a very useful accessory to the telescope. It is attached in place of the eye-piece of the telescope, and is kept pointed to any particular part of the heavens for any length of time necessary for the exposure. This is done by means of a clock-drive, which causes the telescope to move at the same speed as the earth is rotating, but in the opposite direction. If the telescope were not driven in this way there would be trails of the stars on the plates, but by keeping the camera pointed to the same spot in the heavens, long exposures are possible, and much valuable information is obtained.

There are various advantages which photography has over direct vision. One is that the photographic plate shows stars that are only about one-sixth as bright as the faintest that can be seen by the eye using the same telescope. This is because the eye sees all that it can see in an instant, while the plate may be exposed for a long time—an hour or more—thus intensifying the photographic effect. Again, the camera gives a permanent record, which may be studied at leisure by the astronomer. This is especially useful when it is necessary to count the number of stars in a cluster or when there is a programme for mapping the skies.

Photography is invaluable in the discovery of minor planets. As these objects are moving at an appreciable speed, the telescope which is driven to keep a star image in the same position on the plate will not keep an asteroid image fixed. Hence there will be a trail which the astronomer can study later, and in this way he will recognize a new discovery. Fainter novae are discovered by photography, and often a new comet appears on the plate.

The manufacture of plates sensitive to red light and to light far out beyond red in the spectrum has been responsible for many important discoveries in recent times. The invention of faster lenses has also produced valuable results. Modern astronomers would be very badly off without their photographic apparatus.

Questions

1. Describe the general appearance of the Milky Way, from your own observations. Why do you notice rifts and dark parts in its structure?

2. What is the shape of the Milky Way? What appearance do you think it would present to astronomers on planetary systems outside the Milky Way?

3. Name the two nearest extra galactic nebulae to our own galaxy. How far away are they?

4. What is the limit of vision of the largest telescope—the 100-inch at Mount Wilson? When the 200-inch reflector, now under construction, is mounted at Palomar Mountain, U.S.A., what will be its limit?

5. How does the rotation of the Milky Way differ from that of a disc?

6. What is the sun likely to encounter in his journey round the centre of our galaxy? Would this have any effect on the climates of the planets?

7. The distance of some far-off spiral nebulae is about 140 million light-years. With what speed would you expect one of them to be moving?

8. Why do we see some extra galactic nebulae as spirals while others have not this appearance?

9. What is the essential difference between the extra galactic or spiral nebulae and those which belong to our own galaxy? Which are the larger?

10. What instrument is employed to tell us about the chemical elements found in the celestial bodies? In what other special way is it used to supply valuable information about the stars and nebulae?

11. Name some of the advantages of photographic over visual methods of observation.

does not rotate, the centrifugal force at the equator would just balance the attraction of the earth towards its centre and objects would be on the point of being thrown off as the earth rotated

CHAPTER IX

THE ORIGIN OF THE SOLAR SYSTEM

General Orderly Arrangement in the Solar System. It is not a matter of pure chance that all the planets and asteroids describe their orbits in the same direction. The satellites, on the whole, also follow this orderly arrangement, which astronomers call direct orbital motion. There are, however, a few exceptions, which have been referred to in a previous chapter. Among these we may recall the retrograde motion (opposite of direct) of the only satellite possessed by Neptune; the motions of the four satellites of Uranus in the equatorial plane of the planet, which is nearly perpendicular to the plane of the ecliptic, and also the retrograde motion of certain satellites of Jupiter and Saturn. Another important matter which cannot be due to chance is that the planets (but not all the asteroids) move in orbits which are very close to the plane of the ecliptic. All this suggests that there must have been a common cause for the bodies, or at least for most of the bodies, which make up the solar system, and we shall now turn our attention to some of the theories which have been advanced to explain the origin of the planets and satellites.

Laplace's Theory. In 1796 Laplace put forward what is known as the " nebular hypothesis " of the origin of the solar system. He started by assuming a great gaseous mass in rotation, with its matter balanced between the gravitational pull towards its central part and the expansive force of the hot gases. As heat was gradually lost by radiation and dispersed into space, the whole mass would shrink, and hence would rotate more rapidly. This assumption of increasing rotational speed with decreasing size is sound, and actually a rotating and cooling gaseous sphere would spin faster until it bulged out at the equator. Finally, some of the material in the neighbourhood of the equator would be thrown off. We know that if the earth rotated seventeen times as fast as it does at present, the centrifugal force at the equator would just balance the attraction of the earth towards the centre, and objects would be on the point of being thrown off as the earth rotated.

In the case of the rotating gas Laplace assumed that a ring of gaseous matter would be left behind during the shrinking, and the balance would be restored again. The process would be repeated several times, ring after ring being left behind, and each of these rings would eventually form a gaseous ball which would finally become a planet. During the shrinking of these gaseous balls (which themselves would also be rotating) smaller rings would be thrown off, and these would condense into the satellites. On this theory Pluto was formed first of all from the ring which was thrown off when the sun's atmosphere extended out as far as, or beyond, the orbit of this planet, and Mercury was the last planet which was formed. After the ring which was responsible for making Mercury had been left behind, the sun still went on shrinking to about its present size, but no more rings were ejected.

This theory explains a number of features of the solar system, but the present writer has always held that the motions of comets alone were a very strong argument against it. Many people, however, have felt it was so simple, and perhaps so obvious, that it had the appearance of truth, and for a considerable time it was accepted with certain modifications which were necessary in the light of further research. Unfortunately for the theory, it did not satisfy observational evidence, and, in addition, it failed very badly under cross-examination by the dynamical astronomer.

Objections to Laplace's Theory. Among the objections. we may refer to the rings which were supposed to have been left behind during the shrinking of the gas. It was only a pure assumption that rings would be formed; in fact, instead of rings appearing there would be a continuous disc of particles which would not form into planets. Even if a gaseous ring were left behind, it would scatter into space because its own gravitational pull would be too weak to hold it together in the conditions prevailing at the time when the rings were supposed to have been formed.   Again, it has been shown that the sun could never have had sufficient rotational velocity to cause the breaking off of matter in equatorial regions. The theory gives no explanation of both direct and retrograde orbital motion of so many comets; nor does it account for a similar phenomenon which has been observed in the case of several satellites, supposed to have been formed from rings shed by the gaseous ball which formed the planet.   Another difficulty is the inner satellite of Mars, which, as we saw in an earlier chapter, revolves in a shorter time than the planet rotates.   This could not be so if the satellite had been formed from a ring left behind by Mars when it was in a more or less gaseous condition.

FIG. 12.—Showing the tidal effects when two stars approach each other.   The stars are drawn out in egg-shaped fashion. Finally, if they come sufficiently close, disruption occurs.

These and other objections have been urged against the theory, and it is no longer accepted as an explanation of the origin of the planets, satellites, and comets.

The Planetesimal Theory of Chamberlin and Moulton. An interesting theory to explain the origin of the planets was developed in 1900 by Chamberlin and Moulton, the former a geologist and the latter an astronomer, both Americans.   According to this theory a star passed near the sun some thousands of millions of years ago and produced great tides in it (see Fig. 12).   At that time the sun was subject to eruptive forces such as prevail now, but on a more violent scale, and these were intensified by the tidal strains.   The result was that matter was thrown off the surface of the sun, and something like a small spiral nebula was formed.   This idea is worked out in detail by the authors, and it is possible that a visiting star would produce the effects described provided it came sufficiently close. It is assumed that the star came well within the limits of the present planetary system, and though its tidal effects would in themselves be too small to produce disruption unless the centres of the two bodies were only a few millions of miles apart, nevertheless, when these effects were combined with the sun's eruptive forces, something like great ejections of gaseous matter might occur.

When the visiting star had receded, the ejected materials were left revolving around the sun in different orbits, and cooling took place, so that small solid bodies were formed. Larger nuclei then swept up these smaller bodies, and as a result the giant planets were evolved, the asteroids arising from masses of material without any very large nuclei. The satellites grew up around secondary nuclei, which may have been associated with their planets from the time when ejection occurred, or they may have been captured by the planets when they became entangled in the outlying parts of the planetary nuclei.

Professor Bickerton's Theory of Grazing Collision. Before the planetesimal theory was advanced a theory of " grazing collision " was formulated by A. W. Bickerton in New Zealand. Bickerton was a physicist and chemist, but his attention was drawn to astronomy by the appearance of a nova, which he explained by assuming that two bodies had collided, not directly, but with a slanting blow. He applied his " grazing collision " theory to explain many celestial phenomena, and showed how such a collision between a star and the sun could produce the planetary system. His views were published in a New Zealand Journal which was not devoted primarily to astronomy, but later they were embodied in a work with the title, *The Birth of Worlds and Systems*, published in 1911, thirty-three years after he had first propounded his views. He had left New Zealand and had settled down in England shortly before the publication of the work just mentioned, and though his theory was then many years old, it is doubtful if it was known to more than a few British astronomers. Bickerton's enthusiasm in attempting to propagate his tenets regarding novae, the origin of the solar system, and other celestial problems, was never damped by the cold reception which they met in this country. Unfortunately his method of handling some problems in dynamical astronomy left much to be desired, and he gained very few adherents to his theory. It should be added that since his death there are some eminent mathematical astronomers who have not only regarded his theory as feasible, but have also admitted the necessity for a grazing collision to explain the origin of the planets. Although his views were not generally known in this country until thirty years ago, he first formulated them in 1878 and so he antedated the planetesimal hypothesis by many years.

Sir James Jeans's Tidal Theory. Sir James Jeans turned his attention to the problem in 1916 and showed that there was no necessity to assume eruptions on the sun as Chamberlin and Moulton had done. Tidal action produced by the passing star would, he believed, be sufficient to explain disruption of the surface of the sun. He thought that the primitive sun which was disrupted extended to the distances of the outermost planet, but later on he accepted the view of Dr. H. Jeffreys that the sun at the time was something like its present size. If there were a strong central condensation, the approach of another star would not produce disruption into two nearly equal portions like a binary star, but would draw long filaments out of the sun on two sides. The condensation of the filament would ultimately produce the planets, and when these in turn passed near the sun the process would be repeated on a smaller scale, portions of the primitive planets being removed, and these would form the satellites.

Rejection of the Tidal Theory. Although Jeans thought that the visiting star made a very close approach to the sun, he considered it was unnecessary to assume a collision. Dr. Jeffreys showed that certain features in the solar system were inconsistent with the views of Jeans, and he argued that a grazing collision, similar to that which Bickerton suggested, must have taken place. While this assumption rendered it easier to explain some

G

things about the planetary system, other more serious difficulties arose, and the Tidal Theory of Sir James Jeans is no longer accepted as an explanation of the origin of the solar system. The same remark applies to the planetesimal theory. Many other objections have been raised, but it is beyond the range of this elementary treatise to deal with all the objections to both theories, and the reader must accept the statement that they do not explain the origin of the solar system.

Recent Theories. Since the rejection of the Tidal Theory other attempts have been made to explain the planetary system, and these rely on some form of disruption, but with various modifications which, it was believed, would not meet with the usual objections. Dr. R. A. Lyttleton advanced a theory a few years ago which assumed that the sun was a binary star and that its companion was disrupted by the visiting star, the planets and satellites being formed from the debris. This theory was published in the *Monthly Notices of the Royal Astronomical Society* (96, 6, 1936 April), and criticism at different times led to a considerable amount of modification in the original views. Indeed, the theory was modified so much that at the end some portions of it bore very little resemblance to the original, and it has not received general acceptance in astronomical circles. It became very complicated and involved a large number of highly speculative assumptions—not always conducive to a favourable reception. Like some other theories, that of Lyttleton found a difficulty in explaining the origin of the satellites, though other equally formidable problems remained for which no adequate explanation could be found. Lyttleton expressed the views of cosmogonists when he said that nothing approaching exactness can be attained in discussing certain features of the problem.

Another theory was recently published, and in fact is under discussion while this chapter is in course of preparation. In *The Journal of the British Astronomical Association* (52, 6, 1942 July), Mr. J. Miller advanced a theory of the formation of the planets in which a stellar cluster plays a prominent part. The stellar cluster acts like the " visiting star " in some of the earlier theories, except that the sun approaches the cluster, and so is itself the intruder, suffering for this intrusion, in its revolution round some centre of force, by being disrupted. Cohesion of the particles in the sun is an important factor, as it assists the sun's gravitational attraction on the outer layers in preventing disruption until a sufficiently close approach has been made to the cluster. Assuming that cohesion is very small on certain outer layers, these would be the first to be removed when the sun and cluster were a considerable distance apart, and the outermost planet Pluto was formed from the debris. Closer approaches of the sun to the cluster were followed by the ejection of more material as the attraction of the cluster increased through the nearness of the sun and overcame the gravitational pull of the sun on its outer layers and the cohesion of the matter forming these layers. From this ejected material Neptune was formed, and the process was repeated until Mercury was produced from matter ejected at the nearest approach to the central attracting force—the stellar cluster. No attempt is made to explain the origin of the satellites, but it was suggested that a future paper would attack this problem.*

Summary of the Position. The position at the moment is that there is no generally accepted theory of the origin of the planetary system—including, of course, the satellites, comets, and meteors. A tentative view is that somewhere about 3,000 million years ago, or probably longer, matter was ejected from the sun by some means, or perhaps by a combination of circumstances which may not often happen in stellar systems. This matter condensed through its own gravitational attraction into planets of various sizes and densities, Mercury being the least and Jupiter the greatest. Then, when these planets made close approaches to the sun before their orbits became nearly circular, the satellites were torn out of them in turn. In a molten planet the heavier

---

* At the time of reading the proofs, another theory by Mr. B. M. Peek has appeared in *The Journal of the British Astronomical Association* (53, 1, 1942 December). It is a modification of the theory of Laplace.

materials would settle down towards the central regions, and this is what we find in the case of the earth, as explained in Chapter I. In the course of some thousands of years gases would become liquids, and liquids would become solids, but it is difficult to say how long this would require. After the formation of the crust, when cooling had gone far enough and the ocean had condensed, conditions must have become favourable for the development of life.

The process of development and also the necessary conditions still remain a problem, though much has been done in recent years to suggest certain factors that were probably at work. It is possible that life appeared more than 1,000 million years ago, but as it would have been in the form of tiny specks of protoplasm, it could leave no fossil remains. For this reason it is quite impossible to say definitely when life originated, and the few fossils found in the Archaean rocks must represent organisms later than the most primitive forms by many millions of years. There is no reason to suppose that if life developed on other planets it would necessarily evolve along lines similar to those that it has followed on the earth. Here we enter the realm of the biologist, and it is not within our province to discuss this matter further. Perhaps some day the biologist will be able to tell us exactly what life is, but until we know something more about its characteristics we can only speculate on its possible forms on other planets.

Questions

1. Why do you think that the solar system probably originated from one body?

2. Give an outline of Laplace's hypothesis to explain the solar system.

3. From your knowledge of the motions of comets given in Chapter V, why is it difficult to accept Laplace's explanation? What other objections are there to it?

4. What is the essential difference between the planetesimal theory and the tidal theory?

5. If Bickerton's theory of grazing collision is accepted as an explanation or a necessary part of the explanation of the origin of the solar system, why do you think planetary systems must be very rare in the universe?

6. Give a brief outline of some recent theories to explain the origin of the planets and satellites.

APPENDIX

Astronomy and Astrology

In the minds of many people there is a high degree of confusion between astronomy and astrology. The predictions of astrologers are presumed to have a scientific basis, and astronomers are at times invited to cast horoscopes. Some observations on the true relations between the two subjects may therefore be useful.

Astronomy deals with the positions, motions, sizes, and composition of the heavenly bodies, and the present work has, it is hoped, given a clear if brief outline of the scope of this science. Astrology, on the other hand, declares that the heavenly bodies—more especially the planets and the stars near the ecliptic or in the zodiac—have an important influence on the character and destiny of people and on mundane events. It professes to be able to foretell the future of individuals by the positions of the stars at the time of birth, and it even asserts that inanimate objects such as ships, aeroplanes, and trains come under the same influence.

No astronomer of reputation ever concerns himself with astrology, and if he refers to the matter at all it is to express astonishment that belief in it should be so prevalent in this scientific age. As a mathematician he knows that when numerous predictions about events are made, even at random, some of them are bound by the laws of chance to be more or less correct. Among people unfamiliar with the laws of chance such lucky shots make a much deeper impression than the far more frequent cases where astrological predictions are falsified. So the popular belief in astrology tends to survive in spite of repeated exposure of false forecasts. Even the fact that astrologers * failed to predict a major event like the

* Many people do not remember for very long the predictions which appear in the Press, as they do not keep papers in their possession for more than a few days. If readers wish to see how the predictions of astrologers are " fulfilled " they should read

97

outbreak of war in 1939 does not affect the belief among the credulous that astrology is, like astronomy itself, a science.

The astronomer, however, is not specially concerned with the results of predictions, destructive though they are to the scientific reputation of astrology. What interests him most directly is the degree of astronomical knowledge enjoyed by astrologers. If the heavenly bodies play the controlling part that astrologers claim for them, every astrologer ought to be a trained astronomer. Otherwise the very foundation of all his predictions is unreliable.

From this point of view it is instructive to examine the writings of one of the most prominent of modern astrologers. In his book entitled *Complete Practical Astrology*,* Edward Lyndoe asserts that it is possible to proceed with the study of astrology " without so much as the ability to recognize Venus as the bright object shining in the night sky." This assertion is almost equivalent to saying that one may study the art of healing without knowing anything about the human body, and it is not made more convincing by the elementary error of describing Venus as a night star when, as was noted in the chapter on the planets, it is a morning or evening star, and does not shine throughout the night. Again, Jupiter is credited with only nine satellites, although two more were discovered in 1938. Turning to Uranus and Neptune, we are told that the distinguished American astronomer Lowell laid down a programme of research in order to discover the cause of slight discrepancies in the movements of Uranus and Neptune, and that this led to the discovery of Pluto; in point of fact Lowell's determination was based on discordances between the observed and the predicted places of Uranus alone.

---

some of the following : Louis de Wohl, *Commonsense Astrology* (Andrew Bakers, Ltd., 1939). Leonardo Blake, *Hitler's Last Days* (1939) and *The Last Year of the War* (1940), published by the same firm. R. H. Naylor's work, *What the Stars Foretell for 1935* (Hutchinson) will also prove interesting.

* *Complete Practical Astrology*, by Edward Lyndoe. Reprinted 1940. Putnam.

Speaking of the asteroids, this representative astrologer refers specially to Eros, which, he says, "comes nearer the earth than any of the Planets, a few meteors and comets." No astronomer would put meteors in the same class as planets and comets, since we never see meteors until they become luminous by friction with our atmosphere. Apart from that point, the statement quoted is inadmissible. Eros can come within about 14 million miles of the earth, but as far back as 1932 the discovery was made by Delport, in Belgium, of Amor, which comes within 10 million miles. The present writer predicted that other planets coming relatively very close to the earth would eventually be detected. There was nothing astrological about this prediction ; it was based on the probability that out of the scores of thousands of such bodies more than one would come fairly close to the earth. The prediction was fulfilled shortly afterwards by the discovery by Reinmuth of Apollo, which is only 6½ million miles away at its nearest point. Later Adonis and Hermes were added to the list, the former passing a little more than a million miles away and the latter being still less distant. It may be observed that, according to the principles of astrology, all these bodies must have an influence on human affairs. Yet the astrologers of to-day seem to be unaware of their existence, just as their predecessors knew nothing of Uranus or Neptune or Pluto and omitted these planets in their forecasts.

The gulf between astronomy and astrology is made still more apparent when we turn to definitions of familiar terms such as the equinoxes, latitude, and longitude. Mr. Lyndoe defines the equinoxes as follows : "When the sun crosses the equator at the north point of intersection of the ecliptic and equator is the Vernal Equinox, and the first point of Aries ; when the opposite point, is the Autumnal Equinox, and the first point of Libra." In the usual language of science the equinoxes are the two points at which the sun in its apparent course among the stars crosses the celestial equator, and while it is true that the point where the sun's declination changes from south to north is called the Vernal Equinox, such a phrase as " the north point of intersection of the ecliptic and equator " is meaningless to the astronomer.

Latitude, according to the same authority, " whether used for celestial, or Earth, measurements, is the distance north and south of the Equator. On an atlas the lines of latitude are parallel with the Equator—' horizontal '." In reality, however, latitude in the celestial sense has absolutely nothing to do with the equator (it is the angular distance of a body from the ecliptic), and the lines of latitude on a star atlas are not parallel with the equator. A similar confusion is shown in the case of longitude. " Whether used terrestrially or celestially," we are told, longitude " means the distance west or east of the given meridian," and it is added that astronomically it is measured in hours, minutes, and seconds. Actually, celestial longitude has not the remotest connection with the meridian ; it is the angular distance of a body from the first point of Aries, measured on the ecliptic. Moreover, it is measured in degrees, not in hours.

Many more examples of the same sort could be garnered from this and other representative works on astrology, but those already given should suffice to show that astronomers and astrologers move in quite different worlds. It is true that the first astrologers in ancient times were astronomers, in the sense that they observed and measured the movements of the heavenly bodies as accurately as the means at their disposal permitted. The alchemist, likewise, in his search for the Philosopher's Stone, laid the foundations of chemistry, and the witch-doctor, with his magical incantations, opened the way to scientific medicine. The alchemist and the witch-doctor now belong wholly to history and are treated by scientific men as illustrations of the errors into which ignorance and love of the occult may lead us. The astrologer has managed to survive. Various reasons—chiefly psychological—may be given for the continuing popular belief in the cult of astrology, but whatever grounds the devotees may adduce in support of their convictions, they cannot claim the slightest measure of support from astronomers or persuade any scientific man of repute to give astrology the status of a science.

ANSWERS TO QUESTIONS

In many cases it will be sufficient to refer to the pages which describe the problems under consideration.

CHAPTER I

1. Pp. 7–8. 2. P. 8. 3. P. 11. Rotation. 4. (*a*) The axis should be at right angles to the plane of the ecliptic; (*b*) the axis should be in the plane of the ecliptic. 5. About 3 million miles. Not very much. 6. P. 16. Refer to Fig. 2 and the explanation at the end of p. 16. 7. $66\frac{1}{2}°$ north latitude.

CHAPTER II

1. P. 17. 2. Notice her position with regard to the stars from night to night. It will be seen that she has a considerable movement eastward. 3. Make use of Fig. 3 and also of the model or a model similar to that suggested on pp. 19–20. The horns of the moon are the tips of the crescent and the ball representing the moon should show these tips as part of the illuminated surface. The model will show that these tips point away from the torch (the sun) if we imagine that the viewpoint is on the ball representing the earth. 4. P. 22. The chief evidence is based on occultation phenomena. 5. Pp. 23–4. 6. No.

CHAPTER III

1. P. 27. 2. That the outer portions of the sun are in a gaseous condition. 3. P. 29. They are responsible for the discharge of negative charges of electricity which affect our telegraphic services, etc. 4. Seven. 5. An eclipse of the sun. 6. An eclipse of the moon. 7. The most recent determination of the speed of light is 186,271 miles a second, and from these figures it is easily deduced that the times are 8 minutes 11 seconds and 8 minutes

28 seconds, respectively.   8. 21,168 days and 21,888 days, to the nearest day.   9. 2526 followed by ten 0's.

Chapter IV

1. P. 34.   The velocity of escape and the high temperature are two important factors.   2. Owing to the very elliptical orbit, p. 35.   3. P. 35.   4. Carbon dioxide.   Vegetable life (pp. 36–7).   5. P. 36.   The crescent is seen when the planet is in various positions between the earth and the sun, just like the moon.   If Venus is on the side of the sun remote from the earth she appears more like the full moon.   6. Pp. 38–9. Owing to the planet's gravity being small it has probably lost a lot of its atmosphere.   Also, it is possible that most of the oxygen has been taken up by the rocks. 7. Almost certainly natural surface markings (p. 40). 8. Probably to the oxidized rocks (p. 39).   9. Pp. 40–1. 10. The rapid rotation of the planet (p. 41).   11. Probably very extensive, which adds to the apparent diameter of the planet and partly explains its low density in comparison with the earth (p. 41).   12. The spectroscope shows that marsh gas and ammonia exist there and these do not support life such as we know (p. 42).   13. P. 43. 14. It is the only planet which is less dense than water (p. 45).   15. Pp. 46–7.   16. Their plane is turned towards us (p. 46).   17. P. 47.   18. Sir William Herschel (pp. 47–8).   19. They move in orbits which are nearly at right angles to the plane of the ecliptic (p. 48).   20. Life such as we know it could not exist.   21. P. 49.

Chapter V

1. They are very much smaller and also many of them move in orbits which are highly inclined to the ecliptic (p. 51).   2. Because they are too far from Jupiter to have been disrupted by him.   No other planet could have had an effect equal to that of Jupiter (p. 52).   3. Pp. 52–3.   4. P. 56.   In the nucleus (p. 57).   5. Many of them move in orbits highly inclined to the ecliptic.

They differ from the asteroids in the fact that many of them have retrograde motion (p. 55). 6. Light repulsion (p. 56). 7. P. 57. 8. Biela's (p. 58). 9. P. 58. 10. The intense heat caused by friction with the atmosphere produces the glow, but as the bodies are very small they are rapidly burnt up. 11. P. 59. Meteors are more common. 12. Pp. 59–60. 13. Owing to the absence of an atmosphere on the moon there are no weathering effects and hence surface features are more permanent than on the earth.

CHAPTER VI

1. Observations carried out over a few nights will show that a planet changes its position with reference to the stars (p. 61). 2. The base line is too small for their great distances (p. 62). 3. P. 63. 4. P. 65. 5. P. 68. 6. They differ chiefly in their densities (see pp. 66–7). 7. P. 69. 8. (a) Extermination through intense cold or extreme heat, or it will become extinct for reasons not well known, like many other species; (b) extermination through one of the first two conditions. 9. They are so far away that their movements cannot be detected except by the most delicate instruments; hence they are called "fixed stars," and are fixed in so far as we cannot see their movements (p. 62). 10. Nearly 16 times as bright (see pp. 64–5 for method of computation).

CHAPTER VII

1. The former lie nearly in our line of sight but have no physical connection; the latter are revolving around their common centre of gravity (p. 71). 2. No. 3. The mass of the binary system can be found (p. 72). 4. Algol. 5. P. 73. 6. Pp. 74–5. 7. Pp. 75–6. The first theory described is known as the Pulsation Theory; the second is the Theory of Rotational Instability. 8. Its sudden and very great increase in brightness (pp. 76–7). 9. P. 77. The heavenly bodies are too far apart for a sufficient number of collisions to occur to account for all the novae.

Chapter VIII

1. It should be observed at different times of the year and its appearances recorded. Obscuring matter in space cuts off the light in places. 2. Like a bun (p. 79). Its appearance would depend upon their positions. To some it would look like a spiral nebula, but to others in different positions no sign of the spiral structure would be seen (p. 83). 3. P. 80. 4. P. 81. 5. A disc rotates as a solid body, but each star in the Milky Way has its own time of revolution round the centre of gravity of the whole system. 6. Cosmic dust. It is possible (p. 82). 7. Over 13,000 miles a second (see p. 84 for the method of computation). 8. P. 83. 9. P. 84. The extra galactic nebulae. 10. The spectroscope. It is also used to determine line-of-sight velocities of distant stars or nebulae (p. 86). 11. P. 87.

Chapter IX

1. Owing to certain orderly arrangements in the solar system (p. 89). 2. Pp. 89–90. 3. The theory would require that the comets, like the planets, should have direct motion, but about half the comets possess retrograde motion. For other objections see pp. 90–1. 4. The planetesimal theory postulated ejection of matter through eruptions of the sun combined with tidal effects produced by an approaching star, whereas the tidal theory suggested that tidal action alone sufficed to cause the ejection of matter. In addition, according to the planetesimal theory, an enormous number of tiny planets were formed by the ejection, and local condensations swept these up in time, so that they slowly grew into planets. The tidal theory required the formation of the planets quickly from matter which was ejected by the sun. 5. Because of the extreme improbability of collisions taking place. 6. See pp. 94–5.

DIMENSIONS AND DISTANCES OF THE PLANETS FROM THE SUN, AND THEIR DENSITIES AND MASSES

	Diameter in Miles.	Density when Water is the Unit.	Mass when the Earth's Mass is the Unit.	Mean Distance from the Sun in Millions of Miles.	Sidereal Period, or Length of Planet's Year in Years
Sun . .	864,000	1·41	333,434	—	—
Moon . .	2,160	3·34	0·0123	—	—
Mercury .	3,000	3·73	0·037	36·0	0·241
Venus . .	7,600	5·21	0·826	67·3	0·615
Earth . .	[1]7,927, [2]7,900	5·52	1·000	93·0	1·000
Mars . .	4,200	3·94	0·108	141·7	1·881
Jupiter .	[1]88,700, [2]82,800	1·34	318·4	483·9	11·862
Saturn .	[1]75,100, [2]67,200	0·69	95·2	887·1	29·458
Uranus .	30,900	1·36	14·6	1785	84·015
Neptune .	33,000	1·32	17·3	2797	164·788
Pluto . .	?	?	?	3670	247·697

1 and 2 refer to the equatorial and polar diameters, respectively (see pp. 7, 41, and 45).

The computations of the distances of the planets are based on the assumption that the mean distance of the earth from the sun is 93,003,000 miles. This is in accordance with the recent determination of the solar parallax by the Astronomer Royal. As earlier results were based on a mean distance of 92,900,000 miles, there are very slight discrepancies between the figures given above and those found in the text.

The mass of the earth is nearly 6×10^{21} tons, and the mass of any of the other bodies can be found by multiplying this by the figures given in the fourth column. In the case of Pluto, only its distance from the sun and its sidereal period are known with certainty.

NOTES FOR OBSERVERS

AMATEUR astronomers have frequently felt disappointed because their work has seemed fruitless, and in many cases they have given up the study of a subject which had once a great attraction for them. This happens especially with those who work in isolation and have few with whom they can share their interests. Those who intend taking up astronomy seriously are strongly recommended to join some society or association which makes a speciality of the subject. The best Association for amateurs is The British Astronomical Association, which does very valuable work and is always ready, through the Directors of its various Sections, to offer advice to its new members. It holds about nine meetings a year and issues a *Journal* which provides an account of each meeting and contains a number of papers contributed by its members. By joining such an Association it is possible to keep abreast of the times in astronomical development (and the developments in astronomy are both extensive and rapid). Full particulars can be obtained from the Assistant Secretary, 303 Bath Road, Hounslow West, Middlesex. There are fourteen Sections devoted to specialized work in astronomy, the Sun, the Moon, Jupiter, Saturn, Mars, Mercury and Venus, Comets, Meteors, etc., so a member has a wide choice of subjects. Of course there is no obligation to join any particular section; nor is this necessary, because the *Journals* supply the most recent information in astronomical discoveries. The Library contains a fine collection of works on astronomy, and members can borrow the most up-to-date books, which keep them well informed in all branches of the subject.

Star charts are essential if a working knowledge of the constellations and stars is required. A selection of one from the following short list will suffice at first for the beginner:

Philip's Chart of the Stars, by E. O. Tancock.
Sign Posts to the Stars, by F. E. Butler.
The Stars at a Glance.
    Published by George Philip, Fleet St., E.C. 4.

Gall's Easy Guide to the Constellations.
The People's Atlas of the Stars.
    Published by Gall & Inglis, Henrietta St., W.C. 2.

Guide to the Stars, by H. Macpherson.
    Thomas Nelson & Sons, Parkside, Edinburgh.

INDEX